Advanced Structured Materials

Volume 20

Series Editors

Andreas Öchsner
Lucas F. M. da Silva
Holm Altenbach

For further volumes:
http://www.springer.com/series/8611

Frank A. Coutelieris · J. M. P. Q. Delgado

Transport Processes in Porous Media

Springer

Frank A. Coutelieris
Department of Environmental and Natural Resources Management
University of Ioannina
Seferi 2, 30100 Agrinio
Greece

J. M. P. Q. Delgado
Laboratorio de Fisica das Construccoes
Faculdade de Engenharia Universidade do Porto
Rua Dr. Roberto Frias
4200-465 Porto
Portugal

ISSN 1869-8433 e-ISSN 1869-8441
ISBN 978-3-642-27909-6 e-ISBN 978-3-642-27910-2
DOI 10.1007/978-3-642-27910-2
Springer Heidelberg New York Dordrecht London

Library of Congress Control Number: 2011946089

Printed on acid-free paper

Springer is part of Springer Science+Business Media (www.springer.com)

The effort I put for this work is dedicated to my wife, Natasa, who offered me the most valuable gift of my life: our three children.

Frank Coutelieris

Preface

What is a "porous medium"? Although several definitions have been proposed, depending on the application the porous material is involved with, a common sense of this term exists to almost everyone. This is mainly due to their abundance. Either artificial or natural, porous materials can be found in the wide majority of scientific applications, technological developments, and the natural environment. Clothing, chalk, underground oil reservoirs, catalytic pellets and sandstones are typical examples of "everyday" porous materials while other, more modern, approaches consider bread, cooked rice and the human scalp as porous media. Independent of the approach adopted, the basic concepts and properties that characterize a porous medium (porosity, connectivity, consolidation, percolation, tortuosity, homogeneity, isotropy and anisotropy, permeability) have to be well defined and understood before one attempts investigating transport phenomena occurring within them.

"Transport processes in porous media" is a field of great industrial interest, but it still requires more research before it can be properly understood. Fundamental and applied research in heat and mass transfer in porous media has generated increasing interest over the past five decades because of the importance of porous media and transport processes in many engineering applications. A large amount of literature has been generated on this subject and significant advances have been made in modeling fluid flow, heat and mass transfer through porous media including clarification of several important physical phenomena.

The main propose in writing this book is to present a new approach, in several respects. Unlike the most recent reference books on porous media, which catalogue in pure and relatively abstract terms the current position of fundamental research on porous media, this book provides a comprehensive description of the flow and mass transport in porous materials. Authors avoid to present heat transfer processes in porous structures because such a topic is too extensive to be covered along with the other transport processes, better requiring an external book.

The overall approach here is based both on experimental and theoretical state-of-the-art studies, covering the physical and engineering aspects encountered within some of the main applications for porous materials. The text is designed to be intelligible to readers who have some previous knowledge of fundamentals of transport processes as well as a considerable mathematical background. The book's flexible structure makes its accessible to the general engineer or scientist, but also caters for readers with non-scientific backgrounds.

This book was jointly authored from 2010 to 2011, and authors conclude that even if they cannot make any claim of completeness in investigating this crucial field, they believe that some of its recent developments are significantly addressed here Authors would also acknowledge with gratitude the support received from Sandy Coles and Dr. Eleni Vakouftsi as well as current and previous research teams established in the University of Ioannina—Department of Environmental and Natural Resources Management, Greece, the National Center for Scientific Research—Environmental Research Laboratory, Greece, the University of Western Macedonia—Department of Mechanical Engineering, Greece, and the University of Porto—Faculty of Engineering, Portugal. Finally, the authors would welcome reader comments, corrections and suggestions with the aim of improving any future editions.

Frank Coutelieris
J. M. P. Q. Delgado

Contents

Chapter 1
Introduction

Fluid and flow problems in porous media have attracted the attention of industrialists, engineers and scientists from varying disciplines, such as chemical, environmental and mechanical engineering, geothermal physics and food science. Porous medium domains, in which fluid phase mass, phase component mass, momentum, energy, electric charge, and other extensive quantities are being transported while undergoing transformations, are encountered in a variety of disciplines: Civil Engineering (concrete is a porous medium, and soil is a deformable porous medium), Hydrology (aquifers), Environmental Engineering (groundwater pollution by toxic liquids and hazardous wastes), Chemical Engineering (reactors, but the subsurface also behaves like a large chemical reactor in which chemical and biological transformations of contaminants take place), Petroleum Engineering (oil and gas production from reservoirs), Agricultural Engineering (drainage and irrigation), Biomedical Engineering (lungs, kidneys), and Geothermal Engineering (geothermal reservoirs), to mention but a few.

It is widely accepted that most materials are to some extent porous, in the sense that it is quite difficult to find or compose an extremely dense (non-porous) medium. It is evident that the physical and chemical properties of these materials are dependent on their pore structure. For instance, the control of porosity is of great industrial importance in the design of catalysts, industrial adsorbents, membranes and ceramics, as it influences the chemical reactivity of solids and the physical interaction of solids with gases and liquids. The complexity and variety of porous materials has therefore led to a diverse variety of studies on the transport processes occurring there.

1.1 Why Porous Materials?

What is a porous material? The term "porous" is used to describe a complex structure consisting of a compact phase (usually solid) and some void space, where their analogy defines the most significant property of a porous material, its

F. A. Coutelieris and J. M. P. Q. Delgado, *Transport Processes in Porous Media*, Advanced Structured Materials 20, DOI: 10.1007/978-3-642-27910-2_1,

"porosity". Generally speaking, any solid material which contains cavities, channels or interstices may be regarded as porous, although in a particular context a more restrictive definition may be appropriate. Thus, in describing a porous solid, care must be exercised in the choice of terminology in order to avoid ambiguity. More precisely, all these "pores", "cavities" etc. must be carefully defined. Chapter 2 of this book presents a detailed discussion of these terms and their use. For the moment, the following definition sounds accurate enough for the reader: *A porous medium is a region in space comprising of at least two homogeneous material constituents, presenting identifiable interfaces between them in a resolution level, with at least one of the constituent remaining fixed or slightly de-formable* [1].

Why are porous materials interesting? Besides the wide spectrum of applications they have attained during the last centuries, they are not transparent, thus it is necessary to describe any processes occurring within them without local observation, and based only on macroscopic data regarding the inlet and outlet of the domain. The complexity of porous structures allows for assuming very complex local-scale phenomena, which should be integrated to normal macroscopic ones. For instance, even very low volumetric flow-rate would lead to extremely high local velocity and Reynolds number, when the porous structure contains very thin (capillary) throats locally.

The term "porous" is often used instead of the term "dense", which is the most suitable fundamental consideration for the continuum in physics and applied sciences in general. What is important is that dense materials do not actually exist. Even the most compact artificially developed material is of small porosity and must be treated as porous. As a consequence, the study of flow and transport in porous media is applicable to a very wide range of fields, with practical applications in modern industry and environmental areas. In all these disciplines, transport phenomena and transformation occur in porous media domains at a wide range of scales, from the microscopic to the field scale. Understanding these transport phenomena, modeling them at different scales, coping with the uncertainties inherent in such models, and solving the models in order to provide data for management and decision making is the subject of this book.

1.2 Aims and Scopes

This book examines the transport processes that take place in the pore structure of porous materials. Models of pore structure are presented with a discussion of how such models can be used to predict macroscopic quantities that are useful for practical applications. The book is in principal devoted to interpretations of experimental results and simulation predictions in this area, giving directions for future research. Practical applications are given where applicable.

As a first step towards the study of transport processes in porous media, several techniques for artificial representation (modeling) of porous media are discussed. Single phase flows in simplistic and complex porous structures are described in terms of macroscopic and microscopic equations as well as their analytical and numerical solutions. Furthermore, macroscopic quantities such as permeability are introduced and discussed. Mass transport processes in porous media are also discussed and further strengthened by experimental validation and specific technological applications.

The special features of this book are (a) the use of state-of-the-art techniques for modeling transport processes in porous structures, (b) the consideration of realistic sorption mechanisms, (c) the application of advanced mathematical techniques for up-scaling major quantities, and (d) experimental investigation and application, namely, experimental methods for the measurement of relevant transport properties.

1.3 Book Contents and Structure

This book contains seven chapters about transport processes in porous media, organized as follows:

- Initially, Chap. 2 introduces the reader to the *Fundamentals of Porous Structures* by presenting some methods for the description of porous structures. These methods correspond to analytical descriptions (cell models) and numerical ones, either deterministic (digital reconstruction) or probabilistic (stochastic modeling).
- Chapter 3 presents the *Flow in Porous Media*, which is necessary for the adequate estimation of any other transport process. More precisely, this chapter presents the analytical expressions derived in cell models when single phase flow is considered. This is then further exploited to more realistic granular structures.
- *Transport Phenomena in Porous Structures* are presented in Chap. 4, where details of basic phenomena (diffusion and dispersion) are discussed, while a selected application (Dispersion in Packed Beds Flowing by Non-Newtonian Fluids) has been used as an abundant exemplar.
- Chapter 5 deals with the *Modeling of Transport Processes in Porous Materials*, presenting single phase transport in the selected geometries specified in Chap. 3 (cells, granular structures) as well as a novel scale-up technique to obtain macroscopic quantities from microscopic measures/calculations ("Macroscopic quantities for single phase transport").
- The theoretical results of Chap. 5 are further strengthened by the *Experimental Investigation of Transport in Porous Media*, presented in Chap. 6. These experiments contain measurements of molecular diffusion and axial and radial dispersion coefficients, calculation of the solubility of organic compounds at

different temperatures, as well as estimation of tortuosity in porous media. Finally, an application of mass transfer around active solids is presented.

- Five applications of all the above theoretical concepts and experimental results are discussed in Chap. 7 (*Applications and Examples*). These particular applications are: (a) Contaminant Plume Sizes Associated to Different Active Solids, (b) Rising Damp in Building Walls, (c) Bubbles and Drops in Porous Structures, (d) Transport Processes in Solid Oxide Fuel Cells (SOFCs), and (e) Multi Phase Transport in Porous Media.

Each chapter is systematically detailed to be easily understood by a reader with basic knowledge of fluid mechanics, heat and mass transfer, and computational and experimental methods. Although much of the subject matter contained here is available in research literature, journals and conference proceedings, this book presents an overview of the most interesting developments over a wide variety of the most important topics in transport phenomena in porous media. The fundamental concept for this book was to present the complete story of transport processes in porous structures, and for this reason the chapters are self-contained, citing separate references, and presenting their own mathematics, explanations and terminology.

Reference

1. Lage, J.L., Narasimhan, A.: Porous media enhanced forced convection fundamentals and applications. In: Vafai, K. (ed.) Handbook of Porous Media. Marcel Dekker, New York (2000)

Chapter 2
Fundamentals of Porous Structures

2.1 Introduction

The understanding of multiphase flow in porous media is of great importance in many fields such as hydrology, contaminant clean-up and petroleum engineering. Macroscopic properties—principally capillary pressure and relative permeability—are often needed when modelling flow and transport at the continuum scale, whether it is transport of non-aqueous phase liquids (NAPL) in contaminant clean-up or the production of oil during reservoir water flooding. These macroscopic properties are, however, difficult to obtain. It is possible to conduct physical experiments on samples of the reservoir, but this will only reflect one set of conditions. Furthermore, the scale of the reservoir itself is so much larger that a few experiments are unlikely to describe the variation likely to be present. Hence, there is a need to develop physically-based models that can predict multiphase flow and transport properties and their likely variation in a reservoir setting, based on readily available experimental data.

This chapter emphasizes the mathematical modelling of transport processes in porous media by the cell models, the digital reconstruction of the porous structure using the Lattice Boltzmann method and the stochastic modelling to obtain a realistic description of the porous structure.

2.2 Cell Models

Mathematical modeling of transport processes in porous media is a powerful tool employed whenever experimentation is either expensive or difficult due to the nature of the process. On the other hand, the realistic description of the porous structure significantly increases the complexity of the mathematics involved, due to the coupling between the physicochemical mechanisms and the geometrical complexity of the porous medium, thus requiring massive computational power. Given the poor

F. A. Coutelieris and J. M. P. Q. Delgado, *Transport Processes in Porous Media*, Advanced Structured Materials 20, DOI: 10.1007/978-3-642-27910-2_2,

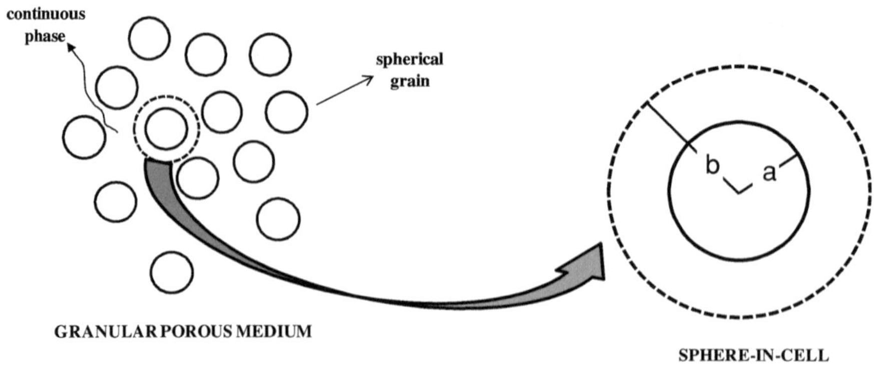

Fig. 2.1 Schematic representation of sphere-in-cell model idealization (a and b denote the radii of the inner and outer spheres, respectively)

computational power available sixty years ago, along with the industry-driven increased necessity for research on porous media structure, in the late 1940s modeling aimed towards the achievement of analytical solutions. The conceptual idea behind this was the cell approach, where the medium is considered as an assemblage of unit cells gathered in a regular manner. Accordingly, it was widely accepted that the unit cell is the adequate representative of the whole medium, therefore processes occurring through the porous structure are described sufficiently by those occurring in the unit cell. As many industrial and technological applications are related with swarms of cylindrical or spherical grains (filtration, separation, fibers, etc.) and the analytical solution for the flow field around an isolated sphere was obtained by Stokes one hundred years ago, what was introduced in those days were the sphere-in-cell models, whose fundamental idea is schematically represented in Fig. 2.1.

Sphere-in-cell models are based on the representation of the overall solid mass of the swarm by a spherical or cylindrical solid body, which is embedded in a spherical or cylindrical liquid envelope, respectively. The boundary conditions imposed on the outer surface of the envelope is supposed to adequately represent the interactions with the other grains of the swarm. Obviously, the thickness of the surrounding fluid layer is adjusted so the ratio of the solid volume to the volume of the liquid envelope to represents exactly the solid volume fraction of the porous medium. The main advantage of these models is that an analytical expression for the stream function can be obtained demanding significantly less effort than that needed for numerical investigations. The spherical shape corresponds to a formulation which leads to axially symmetric flow that has a simple analytical solution of closed form, and can thus be used readily for heat and mass transport calculations. Although this analytical solution is actually an approximation of the real flow field in a complex porous structure, it was sufficient for the engineering applications raised until the early 1980s. On the other hand, the model has one disadvantage in that the outer envelope is not space filling, a difficulty which must be dealt with when a scale-up from the single unit cell to an assemblage of particles is necessary.

The first introduction of sphere-in-cell model was by Cunningham [18]. When considering particle sedimentation, Cunningham postulated that the movement of each spherical particle was allowed only within a concentric mass of fluid boundary. This model had problems in relating the size of the outer fluid sphere to the bed voidage (or particle concentration), therefore Uschida proposed at 1959 an improvement. He assumed that the porous medium is divided into numerous cubes and a particle is placed into the center of each cube. Although space filling, this approach failed to predict accurate flow field solutions due to discontinuities imposed internally by the cubic schema. Richardson and Zaki [59] developed a hexagonal cell model to predict the sedimentation velocity of uniform spheres under viscous flow conditions. In this model, two configurations were considered depending on the vertical particle arrangement. For the first, the distance between neighboring particles in the vertical direction was assumed to be equal to that in the horizontal direction, while the second configuration assumed that vertical neighboring particles are essentially in contact. Due, at least in part, to the oversimplification of the pressure gradient and questionable assumptions regarding the angles between the streamlines and the vertical axis, predictions of the sedimentation velocity from the two configurations were not applicable to a dilute concentration of particles. Happel [26] and Kuwabara [33] presented two independently devised sphere-in-cell models to address the flow problem in sphere packing. The major differences between these two models are in the boundary conditions imposed in the outer surface (A detailed description can be found in Chap. 3). Despite their fundamental differences, both models have been proved to yield very similar flow fields over a wide range of porosity values [67]. At the same time, using formulations analogous to those of spherical geometry, both Happel [26] and Kuwabara [33] also proposed cylinder-in-cell models that consider particles of cylindrical instead of spherical shape. These models were based on the cylindrical-cell model of Kawaguti [32] and found applications in modeling flow through fibrous mats and arrays of parallel fibers. Neale and Nader [45] proposed an important improvement over the aforementioned models. They considered that the basic spherical cell is embedded in an unbounded, continuous, homogeneous and isotropic permeable medium of the same porosity and permeability as those of particle swarm. To describe the flow through the porous surrounding, they used Brinkmans's equation [8]. A decade later, a new spherical cell model for predicting the bed expansion of a liquid–solid fluidized bed was presented by Moritomi et al. [43]. They assumed that Stokes' flow is valid for the fluid within the outer cell boundary while outside the cell the flow is approximated by a potential flow. However, prediction of the bed voidage did not agree well with their experimental data. Another use of Brinkman's equation in cell models is encountered Prasad et al. [54], which replaces the solid sphere of the sphere-in-cell model with a solid sphere surrounded by a concentric spherical shell of homogeneous and isotropic porous material. This model combines features of those developed by Happel and Kuwabara, and uses Brinkman's equation to describe the flow through the porous shell. The work of Prasad et al. is based on the work of Masliyah et al. [42] whose "solid-sphere-with-porous-shell" is embedded in an unbounded fluid instead of the fluid envelope presented by Prasad et al. [54].

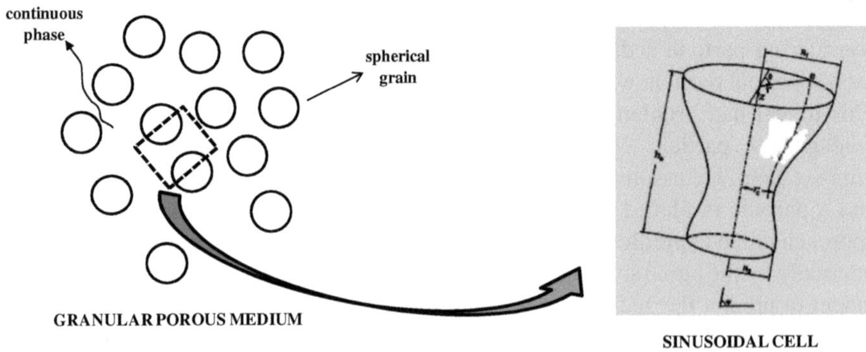

Fig. 2.2 Schematic representation of sinusoidal cell concept

The models described above all considered either spherical or cylindrical grains, which corresponds to two-dimensional spherical geometry after some physical assumptions and mathematical manipulations. However, grain shape is closer to spheroidal than spherical for the majority of practical applications and this is a crucial limitation for model accuracy. Epstein and Masliyah [22] solved numerically the flow-field through clusters of spheroids under creeping flow conditions, whereas the analytical solution of Ammar and Hsieh [4] for Stokes flow inside an oblate hemispheroidal cap was the precursor for the work of Dassios et al. [19, 20]. The latter presented a complete spheroid-in-cell model, quite analogous to those of Happel and Kuwabara, and based on the introduced concept of "semi-separation", an analytical formulation for the stream-function under creeping flow conditions was obtained in terms of series expansions.

Depending on the application considered, several different non grain-type cell models have been proposed and used to simulate the usually complex porous structures. To model the transient behavior of deep-bed filtration systems, Payatakes et al. [50] and Tien et al. [68] proposed the representation of the granular medium by an ensemble of unit-bed elements (UBE), which was found to successfully describe the initial stages of the filtration process. Each UBE is constructed of a number of tube-type cells (collectors), sometimes surrounded by liquid envelopes [58]. Although the trend was to use straight capillaries for the sake of simplicity in the calculations, representation of the pore space by sinusoidal cells, as suggested by Payatakes et al. [51], seemed to produce results in good agreement with experimental observations. The fundamental idea of sinusoidal cells is schematically represented in Fig. 2.2.

The theoretical results of using UBEs for filtration simulations (Payatakes et al. [52]) were found to be in good agreement with experimental data, thus indicating that this simulator type could be used to provide an order-of-magnitude estimate on a truly predictive basis for filtration processes. Some years later, Chiang and Tien [12] used constricted tubes as collectors and employed the concept of UBE in their analysis. Unfortunately, this model cannot describe a complete filter cycle, as it does not deal with the problem of pore clogging and the simultaneous decrease

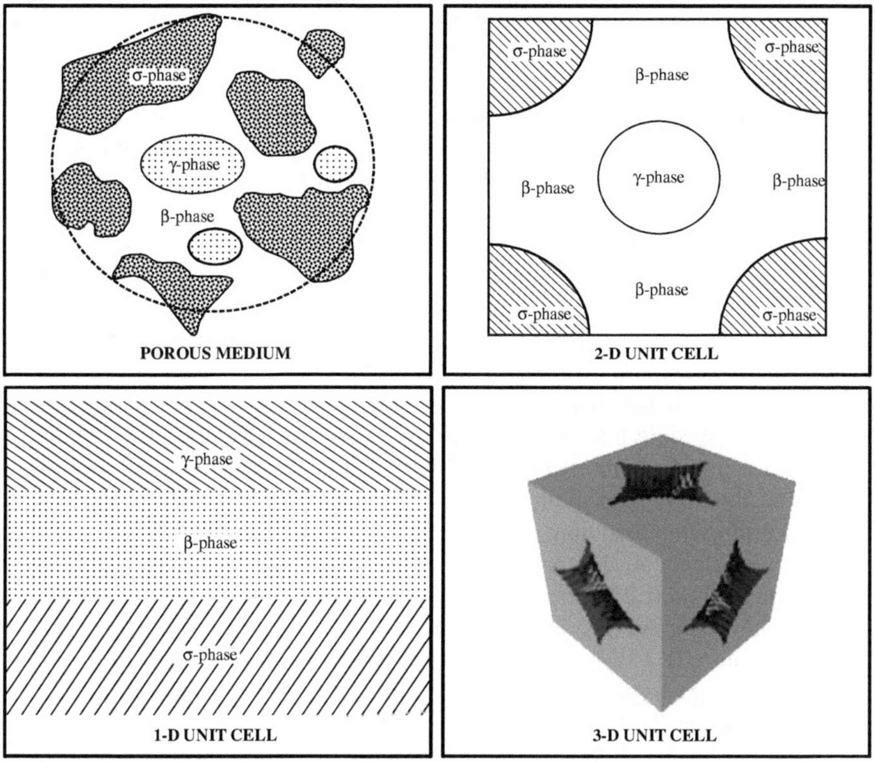

Fig. 2.3 Various unit cells

in permeability. Based on the UBE concept, Choo and Tien [14] presented another approach and assumed that each element contained a number of cylindrical tubes of various sizes. However, it was already known that the assumption of straight capillary tubes oversimplifies pore structure, leading to a substantial underestimation of the particle deposition rate in each cell [49].

Further to representation purposes, unit cells of various dimensions have been introduced and used for the study of diffusion, convection, dispersion and interfacial transport in homogeneous porous media [3, 17, 55–57]. Some of those unit cells are presented in Fig. 2.3.

2.3 Digital Reconstruction (Lattice Boltzmann Method) of the Porous Structure

Transport in porous media is a topic related to many technological and environmental applications. The accurate prediction of the transport coefficients in porous media is a challenging problem due to the complexity of transport mechanisms in fluid–solid systems and the difficulty in representing accurately the complicated and tortuous nature of a porous medium.

The digital reconstruction of porous materials is a relatively new, powerful methodology that enables the reliable representation of the complex biphasic structure induced by porosity. The significant advances in both experimental and numerical practices have allowed the development of reconstruction techniques, the most advantageous being (a) the statistical methods leading to the stochastic reconstruction of porous media in three dimensions based on statistical information obtained from one or several two dimensional images of thin sections of the actual material, and (b) the process-based methods, where the computational procedure tries to imitate the physical processes that commence during the formation of the medium. Although the latter seem to be closer to the physical system, they frequently exhibit severe computational requirements and hence are limited to the specific material considered in each case (Kainourgiakis et al. 31).

The Lattice Boltzmann method (LBM) constitutes a very powerful tool for the study of the hydrodynamical problem of fluid flow inside porous structures, mainly due to the simplified handling of the complicate boundary conditions, as well as, due to the efficiency of the method with regard to parallelization [63, 65]. Lattice Boltzmann method (LBM) is a mesoscopic approach for simulating computational fluid dynamics by solving a discretized Boltzmann equation [11, 60, 71]. An attractive feature of LBM is the ease of addressing complex boundary conditions by implementing very simple schemes. Numerous works have successfully applied LBM in modeling fluid flow in porous media and quantification of porous media permeability [10, 27, 48, 60]. LBM models fluids as particle distributions residing on a discrete lattice, propagating to their adjacent lattice nodes, and colliding with other particles to redistribute momentum.

Aharonov and Rothman [2] first used the Lattice Boltzmann method to simulate the flow of non-Newtonian fluids. Their pioneering work addresses two dimensional pipes and random media. They found that the flux is related to the driving force by a simple scaling law. Similar results were reported by Boek et al. [7]. Recently, Gabbanelli et al. [24] studied the flow of truncated power-law fluids in reentrant flow geometries and found a very good agreement between the results obtained by the Lattice Boltzmann method and those obtained by standard finite element methods, while Sullivan et al. [66] explored the relationship between lattice resolution and simulation accuracy as a function of the power-law index.

2.3.1 Porous Media Generation

Accurate numerical simulation of fluid flow in porous media requires detailed descriptions of porous media morphology, which should include geometric properties such as particle or pore shape and volume, and topological properties such as pore interconnectivity. In many cases, however, the type of model that can be employed is dependent on the modeling method, and more importantly, limited computational resources. It is thus important to construct models that are able to closely mimic the heterogeneity of actual porous media, and at the same time are sufficiently efficient to allow simulation of flow and transport phenomena with reasonable computational effort. In this study, porous media are envisioned as a statistical distribution of non-overlapping circular disks representing soil particles distributed in a rectangular two-dimensional uniform continuum representing the pore space through which a fluid flows.

As first proposed by Gardner, particle size distributions in soil are often assumed to be lognormal in nature (Lerman [36]). Buchan noted that approximately one-half of the US Department of Agriculture textual classification triangle could be adequately modeled by a lognormal distribution. Since a standard lognormal distribution implies zero and infinity for the smallest and largest particle sizes, respectively, modified lognormal distributions were developed to constrain the upper and lower extremes of the particle size. Fredlund et al. [23] proposed a new model based on a unimodal mathematical function, which is believed to provide improved representations of particle size distributions relative to lognormal distributions. This model's ease of use, however, is limited by its employment of five fitting parameters; our study thus employs a modified lognormal distribution to describe particle size distribution, assuming that all particle sizes reside in a 95% confidence interval to eliminate extremely large or small particles.

Li et al. [37] modified the algorithm proposed by Yang et al. [72] for a three dimensional sphere packing, a two-step collective rearrangement technique to generate random porous media. In this work, the particles with size distributions following a modified lognormal distribution are generated until the required porosity is satisfied. The particles are then assigned to a two dimensional domain by assuming a uniform distribution of particle locations. Based on this initial, possibly overlapped configuration (i.e., one particle may overlap another particle), an iterative arrangement process is applied to achieve an overlap free condition. During each iterations the largest particle is selected for relocation if there is any overlap with another particle; if overlap occurs, its spatial location is adjusted until the overlap is removed, and then registered in the final non-overlap location. The procedure continues with the next largest particle, etc. until all particles are registered in their final non-overlap location. Periodic boundary conditions are maintained at all boundaries throughout the iteration process. Figure 2.4 provides an illustration of several of the generated random porous media employed in this study.

2.4 Stochastic Modeling

Most natural and biological phenomena, such as solute transport in porous media, exhibit variability which can not be modeled using deterministic approaches. There is evidence in natural phenomena to suggest that some observations cannot be explained using models which give deterministic solutions. Stochastic processes

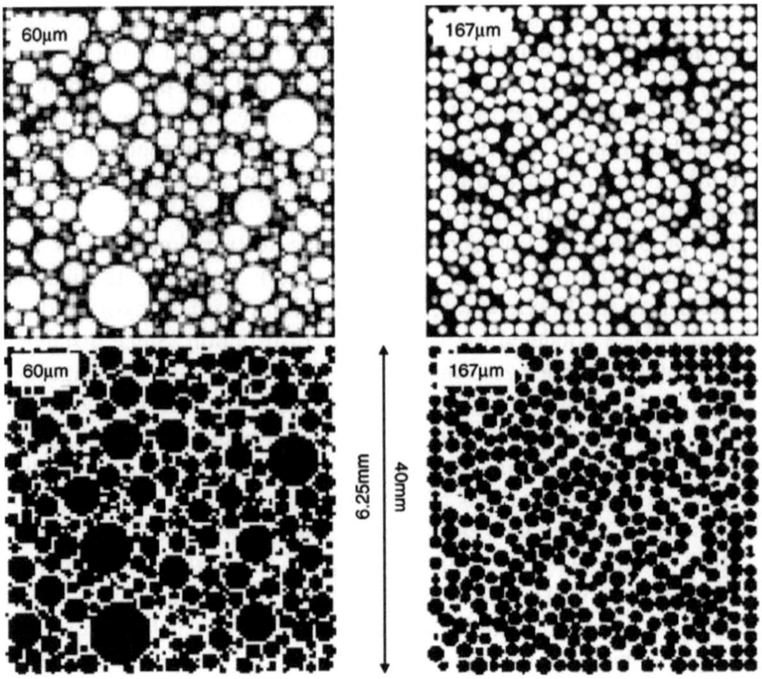

Fig. 2.4 Example of generated porous media

have a rich repository of objects which can be used to express the randomness inherent in the system and the evolution of the system over time.

Transport processes are generally modeled by initial/boundary value problems for a partial differential equation (PDE) describing the local balance of the concentration fields. However, during probabilistic analysis of flow and transport properties in porous media, the uncertainties due to spatial heterogeneity of governing parameters will be often taken into account. The definition of the properties of porous media in space and time using the concept of random functions provides means for (1) studying the inherent heterogeneity, (2) evaluating the spatiotemporal variability of the properties, and (3) assessing the uncertainty associated with their estimated values.

For example, if a small control volume is considered, a mass balance on the solute, in one-dimension, without chemical reaction, leads to

$$\frac{\partial}{\partial x}\left(D_{\mathrm{e}}\frac{\partial C}{\partial x} - uC\right) = \frac{\partial C}{\partial t} \tag{2.1}$$

where D_{e} is the diffusion coefficient, C is the mean solute concentration, u is the interstitial velocity of fluid, and t the time. The velocity field $u(x,t)$ and diffusion coefficients $D_{\mathrm{e}}(x,t)$ instead account for the spatial heterogeneity of the porous media. Within the frame of Scheidegger's theory of dispersion in porous media the components of the diffusion tensor are proportional with those of the velocity field [61]. The velocity field is modelled by the solution of flow equation. For instance, the stationary (in time) and divergence free flow through a saturated porous medium, is governed by Darcy and continuity equations

$$u_i = -\frac{K}{\varepsilon}\frac{\partial H}{\partial x_i}, \quad \sum_{i=1}^{n}\frac{\partial u_i}{\partial x_i} \tag{2.2}$$

where K is the isotropic hydraulic conductivity, ε is the porosity supposed to be constant and H is the hydraulic head. The solution of Eq. 2.2 describes flow through isotropic porous media.

For example, the hydraulic conductivity, K, as a function of x, it is convenient to use its logarithm $Y(x) = \ln K(x)$. This quantity can be inferred from field experiments consisting of pumping tests performed in observation wells. In most experimental settings the locations of the wells are chosen so that an optimum sampling of concentration during tracer tests is obtained. These results in a sparse and non-regular distribution of wells and, therefore interpolation techniques are used to estimate spatial correlations of $Y(x)$. Based on experimentally inferred correlations, stochastic models are proposed. For example, it is assumed that Y is a statistically homogeneous in space random function, normal distributed with constant mean and exponential covariance function [13].

Consequently, Eq. 2.1 has random coefficients. The stochastic approach considers ensembles of solutions of (2.1), corresponding to the ensemble of K field realizations. Expectations and standard deviations of the concentration are computed as ensemble averages. In many applications an upscaled transport model with simpler structure (in general with constant coefficients) is used to describe the process at the desired observation scale. An observable transport process corresponds to a solution of (2.1) for a fixed realization of hydraulic conductivity. Therefore, to obtain realistic predictions, ergodicity assessments have to complete stochastic modelling. Ergodicity can be quantified, for instance, by root mean-square distances between single realization solutions and the output of the upscale model [64]. Technical issues related to the construction of the numerical solutions for PDEs of form (2.1) also provide arguments for a stochastic approach.

2.4.1 The Pore Space

The pore space description can be generated directly using X-ray microtomography (see [16]), where the rock is imaged at resolutions of around a few microns. This can be done due to the differences in X-ray absorption of rock matrix and void space. An example of a 3D image, of Fontainebleau sandstone, is shown in Fig. 2.5a along with a 2D cross section in Fig. 2.6a. This method has the advantage of directly reconstructing the pore space, but due to the specialized scanners required, it is not readily available and also very costly. It is, however,

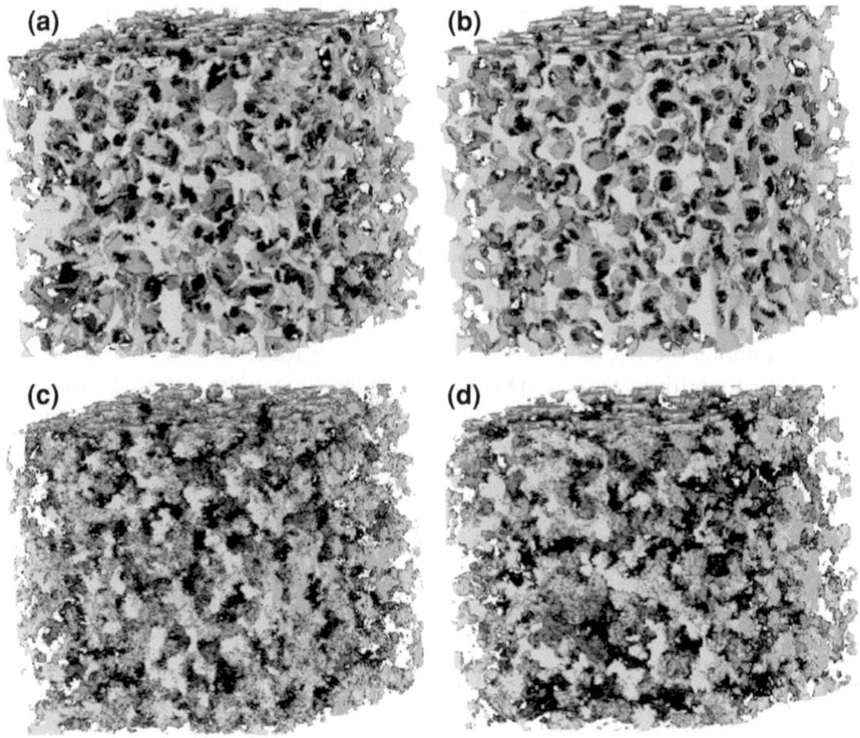

Fig. 2.5 Comparison between different 3D voxel representations of a Fontainebleau sandstone [6], generated using different reconstruction techniques. The side length of each sample is 2.25 mm. **a** X-ray microtomography. **b** Object-based modelling. **c** Gaussian field technique. **d** Simulated annealing

very useful for validating numerical reconstruction techniques of the microstructure, described below.

Most stochastic reconstruction algorithms are based on threshold Gaussian field techniques [1]. These algorithms are based on porosity and two-point correlation functions, both of which can be readily obtained by image analysis of 2D thin sections. The technique is similar to that used in geostatistics. A continuous correlated field is generated using Fourier transform methods and threshold to retrieve the binary phases (pore space and matrix) with the correct porosity and correlation function, as sketched in Figs. 2.5c and 2.6c. This method can also be extended to include more phases, such as clays.

Realising that the earlier Gaussian based methods were not very good at reproducing the underlying particulate structures of the porous media, as evident from Fig. 2.6c, Yeong et al. [73] developed a stochastic method based on simulated annealing, later extended by Manwart et al. [40]. Rather than being restricted to one- and two-point correlation functions, the objective function used can be made to match additional quantities such as multi-point correlation functions,

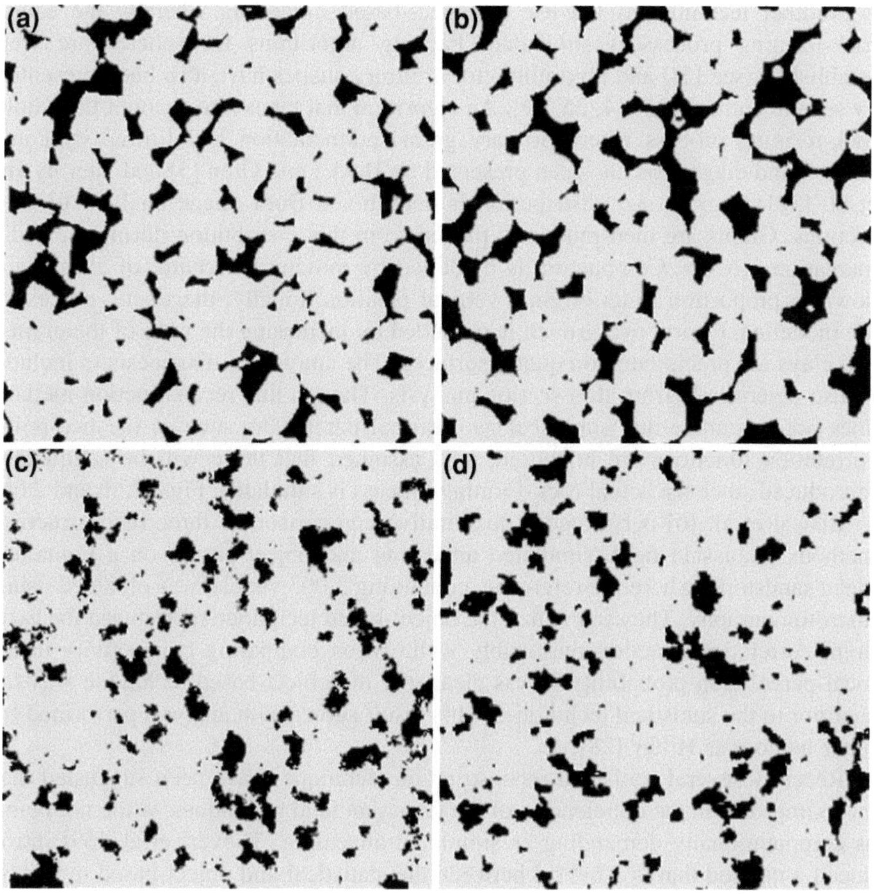

Fig. 2.6 Comparison between different 2D cross-sections of a Fontainebleau sandstone [6], generated using different reconstruction techniques. The side length of each sample is 2.25 mm. **a** X-ray microtomography. **b** Object-based modelling. **c** Gaussian field technique. **d** Simulated annealing

lineal-path function or pore size distribution function to name a few. Since the method is based on moving pore-space voxels around to minimize the objective function, the correct porosity is always retained. They subsequently reconstructed a 3D Fontainebleau and Berea sandstones from 2D sections using an objective function based on both two-point correlation function and either lineal-path or pore size distribution function, Figs. 2.5d and 2.6d. The estimated percolation probability was closer to the reference than when only using the two-point correlation function. Incorporating more higher-order information into the objective function, such as the local percolation probability, would most likely improve the reconstruction further, but that would also increase the computational cost of the method significantly.

Another technique is the use of object-based modelling whereby the actual rock-forming process is simulated. Packing algorithms for spheres are well established (see [1]) and algorithms for arbitrary shapes have also been presented by several authors [15, 34, 35, 53]. An algorithm that takes into account the whole rock-forming process where primary grain sedimentation is followed by compaction and diagenesis has been presented by Bakke and Øren [5] and later by Jin et al. [30]. A grain size distribution is determined from image analysis of thin sections. Grains are then randomly picked from this distribution during the sedimentation process. Compaction is modelled by moving the centre of each grain down in proportion to its original vertical position. Finally, diagenetic processes are modelled. Quartz overgrowth is modelled by increasing the radii of the grains, and clays are precipitated on quartz surfaces. The amount of diagenesis to include is also determined from thin section analysis. Though this reconstruction method does not guarantee that statistical geometrical parameters such as the two-point correlation functions are honoured, it is assumed that these will be adequately reproduced since the actual rock-forming process is simulated, Figs. 2.5b and 2.6b.

Biswal et al. [6] performed a quantitative comparison of three reconstruction methods (Gaussian field, simulated annealing and object-based) on a Fontainebleau sandstone, where the reference, containing 3,003 voxels, was obtained using microtomography. They found that the object-based technique reproduced the two-point correlation function reasonably well. When comparing connectivity using local percolation probability it was clear that the object-based technique was far superior to the statistical techniques. This result agrees with analysis performed by other authors as Hilfer [28].

Recently several statistical reconstruction techniques have been suggested that try to improve on the deficiencies of the Gaussian field techniques, while not being as computationally demanding as simulated annealing. Thovert et al. [69] introduced a method that is a hybrid between the statistical and object-based methods. Their technique is statistical, using only porosity and two-point correlation function, but conditioned to an underlying granular structure, defined using a pore size distribution derived from the two-point correlation function. This is based on a Poissonian penetrable sphere model. They verified their method using a 3D Fontainebleau sample and reported good agreement with the reference, even for the percolation probability. Hilfer and Manswart [29] also developed a hybrid between statistical and object-based method. Initially a close packing of spheres is laid down. Matrix voxels are then randomly distributed in the pore space until a prescribed porosity is attained. The configuration of the added matrix voxels are then updated using simulated annealing, matching the two-point correlation function. They also verified their method using a 3D Fontainebleau sample and the local percolation probability was found to be significantly better than traditional simulated annealing though not quite as good as a traditional object-based method.

For many types of porous media it might, however, be difficult to define the structures that make up the matrix. In carbonate rocks post-depositional diagenesis often completely dominates the matrix structure (see [39]), making object-based techniques difficult to use. In field-scale reservoir characterization the same problems are often experienced where there are multiple faces with distinctive connected geometries; for example, in fluvial reservoirs. Gaussian field techniques will not reproduce the channel connectivity whereas object-based methods rely on the faces geometries being easily parameterized. Strebelle [62] suggested a statistical algorithm in which the multiple-point statistics were inferred from exhaustive 2D training images of equivalent reservoir structures and then used to reconstruct the reservoir, adhering to any conditioning data. This method was applied successfully to both fluvial and more complex patterned reservoirs. The ability to reproduce any pattern makes this method highly attractive for reconstructing complex porous media like carbonates. Okabe et al. [46] have used this algorithm to reconstruct a 3D Fontainebleau sandstone from a 2D training image. Although the granular structure is not as well reproduced as in object-based methods, the local percolation probability is significantly better reproduced than that achieved by other statistical methods such as Gaussian field techniques.

2.4.2 Transport Properties

When estimating single phase transport properties like absolute permeability it is possible to conduct flow experiments directly on the 3D reconstructed sample. Many authors have reported good agreement with experimentally obtained values when solving the Stokes equation for single phase flow using finite difference methods (see [1, 29]) or the Lattice-Boltzmann method [41]. The importance of reproducing long-range connectivity in any reconstruction technique is clearly evident from calculations of absolute permeability. Whereas object-based methods reproduce permeability values within a few percentage points, traditional Gaussian field techniques and simulated annealing typically under-predict permeability by about a factor of ten for low porosity systems that are close to the percolation threshold [29, 41]. For higher porosity systems, the statistical methods generally perform somewhat better at reproducing the local percolation probability [40], with permeability predicted within a factor of about five (see [47]).

Even single phase simulations conducted directly on the 3D reconstructed sample are computationally very expensive. A common way to model both single and multiphase flow quickly on large systems is to represent complex porous media by an equivalent network of pores and connecting throats. Absolute permeability has been successfully predicted, as already mentioned, by Bryant et al. [9] using this method. Another approach is to skip the 3D reconstruction process and estimate absolute permeability directly from the 2D thin sections. Lock et al. [38] computed a distribution of fluid conductance's by recoding pore areas and perimeter lengths. From this a single effective conductance was found using effective medium theory, resulting in a direct estimate of absolute permeability. This approach was applied to a number of sandstone thin sections with predicted absolute permeability's generally within a factor of 2 of the measured values.

For multiphase flow it is possible to use Lattice-Boltzmann techniques (see [11, 25]) to solve the transport equations on the 3D reconstructed sample. Being computationally very demanding, their use in multiphase flow problems is, however, limited to relatively small systems. As a consequence their applications are more tailored towards understanding the fundamental physics of flow in complex systems rather than for quantitative predictions.

A whole range of empirical or semi-empirical methods for predicting relative permeability have suggested over the last five decades (see [21]). One of the most widely used is found by combining the model by van Genuchten [70], relating effective wetting saturation to capillary pressure, with the model for relative permeability by Mualem [44]. Empirical fitting parameters are found by matching the van Genuchten model to experimental capillary pressure. These parameters are subsequently used for the relative permeability predictions. However, what most of these models have in common is that they are only applicable for strongly water-wet data, and many of them also rely on difficult to define parameters such as tortuosity exponents.

References

1. Adler, P.M., Thovert, J.F.: Real porous media: local geometry and macroscopic properties. Appl. Mech. Rev. **51**, 537–585 (1998)
2. Aharonov, E., Rothman, D.H.: Non-newtonian flow (through porous media): a lattice Boltzmann method. Geophys. Res. Lett. **20**, 679–682 (1993)
3. Ahmadi, A., Quintard, M., Whitaker, S.: Transport in chemically and mechanically heterogeneous porous media. Part V: two-equation model for solute transport with adsorption. Adv. Water Resour. **22**, 59–86 (1998)
4. Ammar, L., Hsieh, C.K.: Solution of the velocity and pressure fields for stokes flow inside an oblate hemispheroidal cap. Eur. J. Mech. B/Fluids **10**, 171–178 (1991)
5. Bakke, S., Øren, P.E.: 3-D pore-scale modeling of sandstones and flow simulations in the pore networks. SPE. J. **2**, 136–149 (1997)
6. Biswal, B., Manwart, C., Hilfer, R., Bakke, S., Øren, P.E.: Quantitative analysis of experimental and synthetic microstructures for sedimentary rock. Phys. A: Stat. Mech. Appl. **273**, 452–475 (1999)
7. Boek, E.S., Chin, J., Coveney, P.V.: Lattice Boltzmann simulation of the flow of non-newtonian fluids in porous media. Int. J. Mod. Phys. B **17**, 99–102 (2003)
8. Brinkman, H.C.: A calculation of the viscous force exerted by a bowing fluid on a dense swarm of particles. Appl. Sci. Res. **Al**, 27–34 (1947)
9. Bryant, S.L., Mellor, D.W., Cade, C.A.: Physically representative network models of transport in porous-media. AIChE J. **39**, 387–396 (1993)
10. Cancelliere, A., Chang, C., Foti, E., Rothman, D.H., Succi, S.: The permeability of a random medium: comparison of simulation with theory. Phys. Fluids A: Fluid **12**, 2085–2088 (1990)
11. Chen, S., Doolen, G.D.: Lattice Boltzmann method for fluid flow. Ann. Rev. Fluid. Mech. **30**, 329–364 (1998)
12. Chiang, H.W., Tien, C.: Dynamics of deep bed filtration: analysis of two limiting situations. AIChE J. **31**, 1349–1355 (1985)
13. Chiles, J.P., Delfiner, P.: Geostatisctics: Modeling Spatial Uncertainty. John Wiley and Sons, New York (1999)

14. Choo, C.U., Tien, C.: Simulation of hydrosol deposition in granular media. AIChE J. **41**, 1426–1431 (1995)
15. Coelho, D., Thovert, J.F., Adler, P.M.: Geometrical and transport properties of random packings of spheres and aspherical particles. Phys. Rev. E **55**, 1959–1978 (1997)
16. Coker, D.A., Torquato, S., Dunsmuir, J.H.: Morphology and physical properties of fontainebleau sandstone via a tomographic analysis. J. Geophys. Res-Sol. Ea. **101**, 17497–17506 (1996)
17. Coutelieris, F.A., Kainourgiakis, M.E., Stubos, A.K., Kikkinides, E.S., Yortsos, Y.C.: Multiphase mass transport with partitioning and inter-phase transport in porous media. Chem. Eng. Sci. **61**, 4650–4661 (2006)
18. Cunningham, M.A.: On the velocity of steady fall of spherical particles through fluid medium. In: Proceedings of the Royal Society, London, A83, pp 357–365 (1910)
19. Dassios, G., Hadjinicolaou, M., Payatakes, A.C.: Generalized eigenfunctions and complete semi-separable solutions for stokes flow in spheroidal coordinates. Quart. Appl. Math. **52**, 157–191 (1994)
20. Dassios, G., Hadjinicolaou, M., Coutelieris, F.A., Payatakes, A.C.: Stokes flow in spheroidal particle-in-cell models with Happel and Kuwabara boundary conditions. Int. J. Eng. Sci. **33**, 465–1490 (1995)
21. Dullien, F.A.L.: Porous Media: Fluid Transport and Pore Structure, 2nd edn. Academic, San Diego (1992)
22. Epstein, N., Masliyah, J.H.: Creeping flow through clusters of spheroids and elliptical cylinders. Chem. Eng. J. **3**, 169–175 (1972)
23. Fredlund, M.D., Fredlund, D.G., Wilson, G.W.: An equation to represent grain-size distribution. Can. Geotech. J. **37**, 817–827 (2000)
24. Gabbanelli, S., Drazer, G., Koplik, J.: Lattice Boltzmann method for non-newtonian (power-law) fluids. Phys. Rev. E **72**, 046312–046319 (2005)
25. Gunstensen, A.K., Rothman, D.H.: Lattice-Boltzmann studies of immiscible 2-phase flow through porous-media. J. Geophys. Res-Sol. Ea. **98**, 6431–6441 (1993)
26. Happel, J.: Viscous flow in multiparticle systems: slow motion of fluids relative to beds of spherical particles. AIChE J. **4**, 197–201 (1958)
27. Heijs, A.W.J., Lowe, C.P.: Numerical evaluation of the permeability and Kozeny constant for two types of porous media. Phys. Rev. E **51**, 4346–4352 (1995)
28. Hilfer, R.: Local porosity theory and stochastic reconstruction for porous media. In: Mecke, K.R., Stoyan, D. (eds.) Lecture Notes in Physics, 554th edn. Springer, Berlin (2000)
29. Hilfer, R., Manwart, C.: Permeability and conductivity for reconstruction models of porous media. Phys. Rev. E **64**, 021304–021307 (2001)
30. Jin, G., Patzek, T.W., Silin, D.B.: Physics-based reconstruction of sedimentary rocks. In: Proceedings of the SPE Western Regional/AAPG Pacific Section Joint Meeting, SPE 83587, Society of Petroleum Engineers, Long Beach, USA (2003)
31. Kainourgiakis, M.E., Kikkinides, E.S., Steriotis, T.A., Stubos, A.K., Tzevelekos, K.P., Kanellopoulos, N.K.: Structural and transport properties of alumina porous membranes from process-based and statistical reconstruction techniques. J. Colloid Interf. Sci. **23**, 158–167 (2000)
32. Kawaguti, M.: A hydrodynamical model for the sedimentation. J. Phys. Soc. Jpn. **3**, 209–215 (1958)
33. Kuwabara, S.: The forces experienced by randomly distributed parallel circular cylinder or spheres in a viscous flow at small reynolds numbers. J. Phys. Soc. Jpn. **14**, 527–531 (1959)
34. Latham, J.P., Lu, Y., Munjiza, A.: A random method for simulating loose packs of angular particles using tetrahedral. Geotechnique **51**, 871–879 (2001)
35. Latham, J.P., Munjiza, A., Lu, Y.: On the prediction of void porosity and packing of rock particulates. Powder Techn. **125**, 10–27 (2002)
36. Lerman, A.: Geochemical Processes: Water and Sediment Environments. Wiley, New York (1976)

37. Li, Y., LeBoeuf, E.J., Basu, P.K., Mahadevan, S.: Stochastic modeling of the permeability of randomly generated porous media. Adv. Water Resour. **28**, 835–844 (2005)
38. Lock, P.A., Jing, X.D., Zimmerman, R.W., Schlueter, E.M.: Predicting the permeability of sandstone from image analysis of pore structure. J. Appl. Phys. **92**, 6311–6319 (2002)
39. Lucia, F.J.: Carbonate Reservoir Characterization. Springer, Berlin (1999)
40. Manwart, C., Torquato, S., Hilfer, R.: Stochastic reconstruction of sandstones. Phys. Rev. E **62**, 893–899 (2000)
41. Manwart, C., Aaltosalmi, U., Koponen, A., Hilfer, R., Timonen, J.: Lattice-Boltzmann and finite-difference simulations for the permeability for three dimensional porous media. Phys. Rev. E **66**, 016702–016713 (2002)
42. Masliyah, J.H., Neale, G., Malysa, K., Van De Ven, G.M.: Creeping flow over a composite sphere: solid core with porous shell. Chem. Eng. Sci. **42**, 245–253 (1987)
43. Moritomi, H., Yamagishi, T., Chiba, T.: Prediction of complete mixing of liquid-fluidized binary solid particles. Chem. Eng. Sci. **41**, 297–305 (1986)
44. Mualem, Y.: A new model for predicting the hydraulic conductivity of unsaturated porous media. Water Resour. Res. **12**, 513–522 (1976)
45. Neale, G.H., Nader, W.: Prediction of transport processes within porous media: creeping flow relative to a fixed swarm of spherical particles. AIChE J. **20**, 530–538 (1974)
46. Okabe, H., Blunt, M.J.: Pore space reconstruction using multiple-point statistics. J. Petrol. Sci. Eng. **46**, 121–137 (2005)
47. Øren, P.E., Bakke, S.: Reconstruction of berea sandstone and pore-scale modeling of wettability effects. J. Petrol. Sci. Eng. **39**, 177–199 (2003)
48. Pan, C., Hilpert, M., Miller, C.T.: Pore-scale modeling of saturated permeabilities in random sphere packings. Phys. Rev. E **64**, 0066702-1–0066702-9 (2001)
49. Paraskeva, C.A., Burganos, V.N., Payatakes, A.C.: Three-dimensional trajectory analysis of particle deposition in constricted tubes. Chem. Eng. Commun. **108**, 23–48 (1991)
50. Payatakes, A.C., Tien, C., Turian, R.M.: A new model for granular porous media. Part I: model formulation. AIChE J. **19**, 58–67 (1973)
51. Payatakes, A.C., Tien, C., Turian, R.M.: Trajectory calculation of particle deposition in deep bed filtration. Part I: model formulation. AIChE J. **20**, 889–900 (1974)
52. Payatakes, A.C., Brown, D.H., Tien, C.: On the transient behavior of deep bed filtration. In: Proceedings of the 83rd National AIChE Meeting, Houston, USA (1977)
53. Pilotti, M.: Reconstruction of clastic porous media. Transport. Porous Med. **41**, 359–364 (2000)
54. Prasad, D., Narayan, K.A., Chhabra, R.P.: Creeping fluid flow relative to an assemblage of composite spheres. Int. J. Eng. Sci. **28**, 215–230 (1990)
55. Quintard, M., Whitaker, S.: One- and two-equation models for transient diffusion processes in two-phase systems. Adv. Heat Transf. **23**, 369–464 (1993)
56. Quintard, M., Whitaker, S.: Convection, dispersion, and interfacial transport of contaminants: homogeneous porous media. Adv. Water Resour. **17**, 221–239 (1994)
57. Quintard, M., Whitaker, S.: Transport in chemically and mechanically heterogeneous porous media I: theoretical development of region-averaged equations for slightly compressible single-phase flow. Adv. Water Resour. **19**, 29–47 (1996)
58. Rajagopalan, R., Tien, C.: Trajectory analysis of deep bed filtration using the sphere-in-cell model. AIChE J. **22**, 523–528 (1976)
59. Richardson, J.F., Zaki, W.N.: The sedimentation of a suspension of uniform spheres under conditions of viscous flow. Chem. Eng. Sci. **3**, 65–73 (1954)
60. Rothman, D.H.: Cellular-automaton fluids: a model for flow in porous media. Geophysics **53**, 509–518 (1988)
61. Scheidegger, A.E.: General theory of dispersion in porous media. J. Geophys. Res. **66**, 3273–3278 (1961)
62. Strebelle, S.: Conditional simulation of complex geological structures using multiple-point statistics. Math. Geol. **34**, 1–21 (2002)

63. Succi, S.: The Lattice Boltzmann Equation for Fluid Dynamics and Beyond. Oxford University Press, Oxford (2001)
64. Suciu, N., Vamos, C., Vanderborght, J., Hardelauf, H., Vereecken, H.: Numerical investigations on ergodicity of solute transport in heterogeneous aquifers. Water Resour. Res. **42**, W04409–W04419 (2006)
65. Sukop, M.C., Thorne Jr, D.T.: Lattice Boltzmann Modeling an Introduction for Geoscientists and Engineers. Springer, Berlin (2006)
66. Sullivan, S.P., Gladden, L.F., Johns, M.L.: Simulation of power-law fluid flow through porous media using lattice Boltzmann techniques. J. Non-Newton. Fluid Mech. **133**, 91–98 (2006)
67. Tien, C.: Granular Filtration of Aerosols and Hydrosols. Butterworths, USA (1989)
68. Tien, C., Turian, R.M., Pendse, H.P.: Simulation of the dynamic behavior of deep bed filters. AIChE J. **25**, 385–395 (1979)
69. Thovert, J.F., Yousefian, F., Spanne, P., Jacquin, C.G., Adler, P.M.: Grain reconstruction of porous media: application to a low-porosity fontainebleau sandstone. Phys. Rev. E **63**, 061307–061323 (2001)
70. Van Genuchten, M.T.: A closed-form equation for predicting the hydraulic conductivity of unsaturated soils. Soil Sci. Soc. Am. J. **44**, 892–898 (1980)
71. Wolf-Gladrow, D.A.: Lattice-Gas Cellular Automata and Lattice Boltzmann Models: An Introduction. Lecture Notes in Mathematics. Springer, New York (2000)
72. Yang, A., Miller, C.T., Turcoliver, L.D.: Simulation of correlated and uncorrelated packing of random size spheres. Phys. Rev. E **53**, 1516–1524 (1996)
73. Yeong, C.L.Y., Torquato, S.: Reconstructing random media. Phys. Rev. E **57**, 495–506 (1998)

Chapter 3
Flow in Porous Media

3.1 Introduction

As mentioned in Chap. 1, a porous medium is a particulate phase (usually solid) that contains void spaces (microscopic pores). These pores may either be connected to each other or unconnected, and are distributed in the medium in either a regular or a random manner. Porous media can be distinguished as *granular* or *fractured* in form with either *consolidated* or *unconsolidated* mechanical properties. In a consolidated porous medium, the particles (grains) are connected by an intermediate cementing material, while in an unconsolidated porous medium the grains are loose.

The spatial distribution of matter in a porous medium can be typically represented by the phase function $Z(\mathbf{x})$, defined as:

$$Z(\mathbf{x}) = \begin{cases} 1 & \mathbf{x} \text{ belongs to the pore space} \\ 0 & \text{otherwise} \end{cases} \tag{3.1}$$

where $\mathbf{x}$ is the position vector from an arbitrary origin.

The motion of a continuum is generally described by a usually linear relation between some fluxes and the relative driving forces. Following the trend of *determinism*, a large number of constitutive (phenomenological) equations describing relationships between fluxes and driving forces exist in several fields of physics. These include Newton's law (which correlates developing forces with acceleration, i.e. velocity gradient), Fourier's law (which correlates heat flow with temperature gradient), Fick's law (which correlates mass flow with concentration gradient), Ohm's law (which correlates current with potential gradient), etc. The description of flow in porous media is especially difficult due to the media's complex geometry. Furthermore, a general law of the continuum theory is the law of conservation of extensive properties such as mass, momentum and energy. The resulting equations of continuity are commonly referred to as field equations and must be considered along with flow equations to adequately describe flow in porous structures.

F. A. Coutelieris and J. M. P. Q. Delgado, *Transport Processes in Porous Media*,
Advanced Structured Materials 20, DOI: 10.1007/978-3-642-27910-2_3,

Experimental and theoretical investigations are normally associated with the scale of the application considered. In general, these scales are microscopic (or "pore" scale), mesoscopic (or "local" scale), and macroscopic (or "field" scale). As far as the phenomena considered (and the governing equations as well as the relative solutions) are strongly dependent on the scale, the transition between different scales is of great importance. This transition is usually considered from the lowest "pore" level, where experimental and simulation results could be easily obtained, to the higher "field" scale, where data are hard to be measured. This up-scaling process is the subject of many studies (see Sects. 5.5 and 7.5).

3.1.1 Macroscopic Description

During his experiments on flow in pipes, Darcy found and proved that the pressure drop caused by the flow is proportional to the velocity. This result is expressed by the phenomenological Darcy law equation [7]:

$$q = A \frac{k}{\mu} \frac{\Delta p}{L} \tag{3.2}$$

which generally describes the fluid flow through a porous medium. In the above formula, q is the volumetric flowrate through a cross-section, A of the porous medium, which is perpendicular to the flow direction, L is the length of the porous media in direction of flow, Δp is the pressure difference along the porous medium, μ is the viscosity of the flowing fluid, and k is the permeability as a material property of the porous medium. This equation can be written in differential form as:

$$\nabla p = -\frac{\mu}{k} \mathbf{u} \tag{3.3}$$

where $\mathbf{u}$ denotes the velocity vector of the fluid. It is important to note that the Darcy law is only valid for a laminar and steady state one-phase flow through a porous medium. In addition, the fluid must be largely incompressible.

An extension to the Darcy law is the isotropic law introduced by Brinkman law [3]:

$$\nabla p = -\frac{\mu}{k} \mathbf{u} + \mu_{eff} \nabla^2 \mathbf{u} \tag{3.4}$$

where μ_{eff} is an effective viscosity which may be different from μ. The Brinkman law is a slight modification of the Stokes equation and one of its practical advantages is that applications and analytical tools devoted to the Stokes approach can be used with small adaptations.

One of the most significant applications of Darcy's law (3.2) is the calculation of permeability of a porous material. After solving the equation for k:

$$k = \frac{q}{A} \mu \frac{L}{\Delta p} \tag{3.5}$$

the permeability of a porous media is defined as 1 Darcy, if a fluid of 1 cP viscosity, flowing through a cross-section of 1 cm^2 at a rate of 1 cm^3/s, causes a pressure drop of 1 atm/cm. Many years later, Klinkenberg observed that gas permeability is not the same as liquid permeability for the same porous structure, since gas permeability depends on pressure [16]. The correlation between gas permeability k_{gas}, liquid permeability k_{liq} and mean pressure inside the core p_{mean}, is given as:

$$k_{gas} = k_{liq}\left(1 + \frac{b}{p_{mean}}\right) \tag{3.6}$$

where parameter b depends on the gas studied.

All the above calculations are valid under the major restriction that permeability should be a number, which means that the porous medium is homogeneous and isotropic. For non-homogeneous and anisotropic media, permeability becomes a *tensor*, thus the above equations should be transformed accordingly.

3.1.2 Microscopic Description

Modeling approaches for microscopic conditions can be described by the appropriate flow conditions equations (creeping, laminar, turbulent, etc.). It is important to note that these equations are defined in the pore space, therefore the boundary conditions in the solid–fluid interface are essential to solve the flow problem. Obviously, equations of motion and the accompanying boundary conditions, are strongly dependent on the application considered, thus many different approaches have been proposed in the literature. The problem here is the derivation of representative macroscopic quantities *from* the microscopic results, a procedure which hardly allows for a generalized unique solution. As the extensive discussion of these issues is not within the scope of this book, our presentation is limited to *granular porous media*, where analytical solutions are also available through the cell models.

3.2 Analytical Solutions for Single Phase Flow in Cell Models

As mentioned in Chap. 2, cell models have been widely used for mathematical simulation of the flow conditions through porous structures, especially granular materials. According to these models, the grains are distributed throughout the predefined space, and each grain is enclosed in a cell formed by the liquid phase. The cell model can be used to reduce the solution of the boundary-value problem for the flow around a system of particles to the problem for a single particle, where the flow field solution can be obtained by the analytical solution of the Stokes problem. The cell models differ in their boundary conditions, as discussed below. A wide variety of assumptions exists for the shape of the grains, and the cells in general: spherical, cylindrical, spheroidal, etc. Different approaches for the physical conditions and mathematical formulations have also been presented for each geometry considered, therefore producing an extensive range of combinations between geometrical and physical considerations. These combinations are described below.

3.2.1 Sphere-in-Cell Models: Kuwabara's Approach

Consider a solid sphere of radius α, surrounded by another concentric spherical liquid envelope of radius β, whose thickness is adjusted so the porosity of the medium is equal to that of the model. The internal sphere is solid and stationary while a Newtonian fluid flows around the solid core under constant approaching velocity. The governing equation for quasi-steady creeping flow of incompressible (i.e. constant density) viscous (i.e. constant dynamic viscosity) fluids is the well-known Stokes equation, given as:

$$\nabla p = \mu \nabla^2 \mathbf{v} \tag{3.7}$$

along with the continuity equation:

$$\nabla \cdot \mathbf{v} = 0 \tag{3.8}$$

where p is the pressure field, $\mathbf{v}$ is the velocity vector, and μ is the dynamic viscosity. As the system is axially symmetrical, the problem is two-dimensional. Accordingly, by using the spherical coordinates system (r, θ), the above equation can be written in terms of the steam-function ψ, as:

$$\mathrm{E}^2\left(\mathrm{E}^2 \psi\right) = 0 \tag{3.9}$$

where:

$$\mathrm{E}^2 = \frac{\vartheta^2}{\vartheta r^2} - \frac{\sin\theta}{r^2}\frac{\vartheta^2}{\vartheta \theta^2} \tag{3.10}$$

Consequently, the velocity components are given as:

$$u_r = -\frac{1}{r^2 \sin\theta}\frac{\vartheta \psi}{\vartheta \theta}, \quad u_\theta = \frac{1}{r \sin\theta}\frac{\vartheta \psi}{\vartheta r} \tag{3.11}$$

To solve the above problem, it is necessary to consider the appropriate boundary conditions. According to Kuwabara's approach [17], zero vorticity is assumed on the outer surface, as depicted in Fig. 3.1.

The velocity components are expressed as:

$$u_r = -2\left[\frac{F_1}{r^3} + \frac{F_2}{r} + F_3 + F_4 r^2\right] \cos\theta \tag{3.12}$$

$$u_\theta = -\left[\frac{F_1}{r^3} - \frac{F_2}{r} - 2F_3 - 4F_4 r^2\right] \sin\theta \tag{3.13}$$

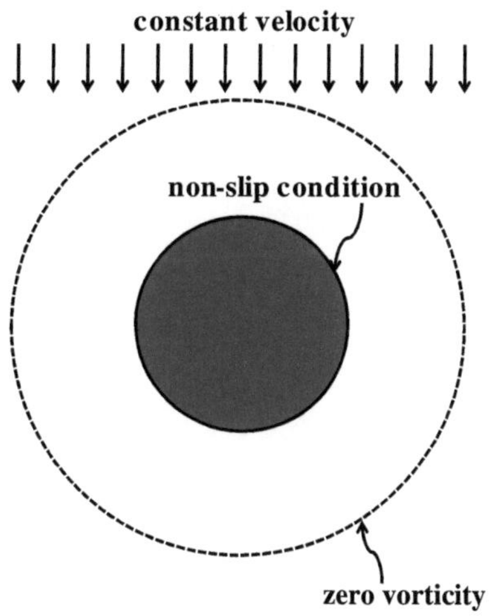

Fig. 3.1 Kuwabara's sphere-in-cell model

where:

$$F_1 = -\frac{U_\infty}{4F_5}\left(1 - 2\frac{\alpha^3}{\beta^3}\right) \tag{3.14}$$

$$F_2 = \frac{3U_\infty}{4F_5} \tag{3.15}$$

$$F_3 = -\frac{U_\infty}{2F_5}\left(1 + \frac{\alpha^3}{2\beta^3}\right) \tag{3.16}$$

$$F_4 = -\frac{3U_\infty}{20F_5}\left(\frac{\beta^3}{\alpha^3}\right) \tag{3.17}$$

$$F_5 = \left(1 - \frac{\alpha}{\beta}\right)^3\left(1 + \frac{6\beta}{5\alpha} + \frac{3\beta^2}{5\alpha^2} + \frac{\beta^3}{5\alpha^3}\right) \tag{3.18}$$

and where U_∞ is the magnitude of the uniform approaching velocity.

Typical results for the stream function are presented in Fig. 3.2 which also depicts a comparison with Happel's approach (for details on Happel's model, see Sect. 3.2.2).

3.2.2 *Sphere-in-Cell Models: Happel's Approach*

The model proposed by Happel [10, 11] is similar to that of Kuwabara but differs in the fluid motion. Kuwabara's stationary grain is located in a flowing fluid, however, Happel proposed a grain moving under constant velocity in an otherwise

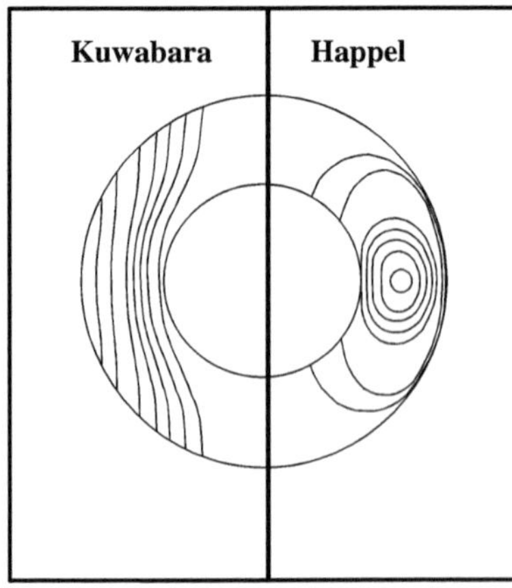

Fig. 3.2 Sample stream lines in two sphere-in-cell models

quiescent surrounding fluid. This dissimilarity imposes different boundary conditions, as depicted in Fig. 3.3.

Compared to Kuwabara's approach, Happel's model has the additional advantage of being autonomous from an energy point of view. This benefit is the main reason for the wider acceptance of Happel's approach rather than Kuwabara's model.

In Happel's approach, the governing equations are the same as (3.7–3.11), where it is assumed that the solid spherical core is moving under a constant velocity gradient q. After all, the velocity components are expressed as [10]:

$$u_r = \left[6r^3 F_1 + 2rF_2 + \frac{6}{r^2}F_3 - \frac{3}{r^4}F_4 + rq\right]\sin\theta\cos\theta \tag{3.19}$$

$$u_\theta = \left[5r^3 F_1 + rF_2 + \frac{1}{r^4}F_4 + \frac{rq}{2}\right]\left(\cos^2\theta - \sin^2\theta\right) \tag{3.20}$$

where:

$$F_1 = -\frac{5q}{4\alpha^2}\left(\frac{\left(\frac{\alpha}{\beta}\right)^7}{10 + 4\left(\frac{\alpha}{\beta}\right)^7}\right)F_5 \tag{3.21}$$

$$F_2 = \frac{5q}{4}\left(\frac{4 + 10\left(\frac{\alpha}{\beta}\right)^7}{10 + 4\left(\frac{\alpha}{\beta}\right)^7}\right)F_5 - \frac{q}{2} \tag{3.22}$$

$$F_3 = -\frac{5q\alpha^3}{12}F_5 \tag{3.23}$$

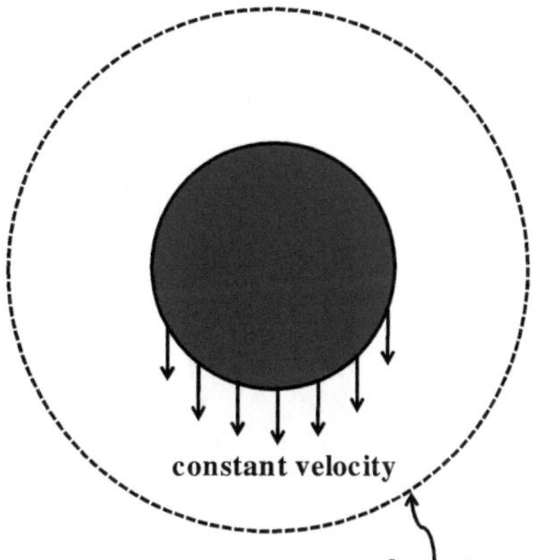

Fig. 3.3 Happel's sphere-in-cell model

$$F_4 = -\frac{5\alpha^5}{10 + 4\left(\frac{\alpha}{\beta}\right)^7} F_5 \tag{3.24}$$

$$F_5 = \frac{10 + 4\left(\frac{\alpha}{\beta}\right)^7}{10\left[1 - \left(\frac{\alpha}{\beta}\right)^{10}\right] - 25\left(\frac{\alpha}{\beta}\right)^3\left[1 - \left(\frac{\alpha}{\beta}\right)^4\right]} \tag{3.25}$$

The visual interpretation of sample flow in Happel's sphere-in-cell is depicted in Fig. 3.2.

3.2.3 Cylinder-in-Cell Models

Further to "sphere-in-cell" models, both Happel and Kuwabara also proposed similar "cylinder-in-cell" ones. The similarity in the formulations and the consequent analytical solutions is due to the two-dimensionality of the mathematical description that allows us to ignore the shape imposed by the third dimension.

3.2.4 Spheroid-in-Cell Model

As described above, grain are often closer in shape to spheroids rather than spheres. This observation leads to the development of "spheroidal-in-cell" models, similar to those of Happel and Kuwabara. One difficulty of such a mathematical interpretation arises from the orientation of the spheroidal geometry, which can be either *prolate* or *oblate*, as illustrated in Fig. 3.4.

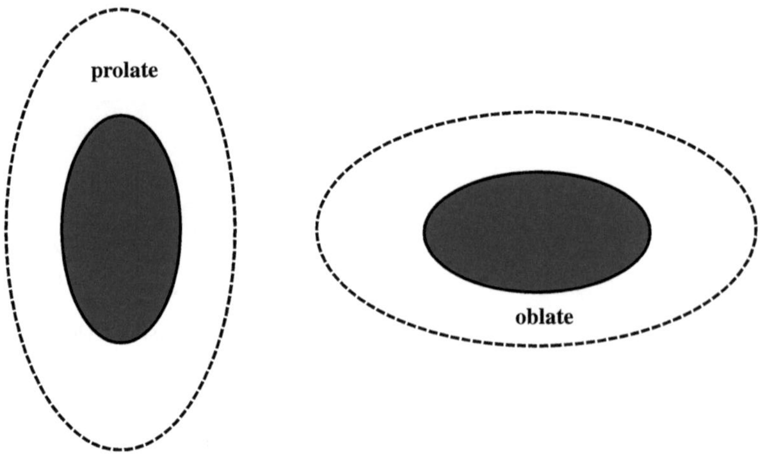

Fig. 3.4 Prolate and oblate spheroidal cells

In cases of prolate geometry, the inner solid spheroid has long semiaxis a_3 and short semiaxis a_1, where $a_3 > a_1$. The semifocal distance a is defined as $a = \sqrt{a_3^2 - a_1^2}$. The outer confocal prolate spheroid is of long and short semiaxes b_3 and b_1, respectively, while its dimensions are determined so that the volume fraction of the spheroid-in-cell is equal to that of the original swarm of spheroidal particles. Thus, given the dimensions of the inner spheroid, porosity is related to the outer semiaxes by the following equations:

$$(1 - \varepsilon) b_1^2 b_3 = a_3 \tag{3.26}$$

$$b_1 = \sqrt{b_3^2 - a^2} \tag{3.27}$$

The axisymmetric, steady state, incompressible, viscous creeping flow is again described in terms of stream function by Eq. 3.9, where the operator E^2 is given as:

$$E^2 = \frac{1}{\alpha^2 \left(\sinh^2 \eta + \sin^2 \theta\right)} \left[\coth^2 \eta \frac{\partial^2}{\partial \eta^2} - \coth^2 \eta \frac{\partial}{\partial \eta} + \frac{\partial^2}{\partial \theta^2} - \cot \theta \frac{\partial}{\partial \theta}\right] \tag{3.28}$$

Under this respect, the velocity components are given in the prolate spheroidal coordinates system (η, θ) as [20]:

$$u_\eta = \frac{-1}{\alpha^2 \sqrt{\sinh^2 \eta + \sin^2 \theta} \sinh \eta \sin \theta} \frac{\partial \psi}{\partial \theta} \tag{3.29}$$

$$u_\theta = \frac{1}{\alpha^2 \sqrt{\sinh^2 \eta + \sin^2 \theta} \sinh \eta \sin \theta} \frac{\partial \psi}{\partial \eta} \tag{3.30}$$

By using the *semi-separation concept* introduced by Dassios et al. [8], the stream function is analytically given by an infinite series expansion of the form:

$$\psi(\eta,\theta)=\sum_{n=2,4,\ldots}^{\infty} g_n(\cosh\eta)G_n(\cos\theta) \tag{3.31}$$

The leading term of the above infinite series has proved to be sufficient enough to adequately represent the whole stream function [4], thus the stream function is given as:

$$\begin{aligned}\psi(\eta,\theta)=\{&A_1G_1(\cosh\eta)+A_2G_2(\cosh\eta)+A_3G_4(\cosh\eta)\\&+A_4H_2(\cosh\eta)\}G_2(\cos\theta)\end{aligned} \tag{3.32}$$

where A_1, A_2, A_3 and A_4 are constants coefficients, the values of which depend on the geometrical parameters η_a and η_β (the values of the coordinate η on the inner and outer surface, respectively). Obviously, their expressions are also dependent on the model considered (Kuwabara or Happel), i.e. on the boundary conditions applied.

More specifically, Kuwabara's approach corresponds to the expression:

$$\begin{aligned}\psi(\eta,\theta)=\frac{\alpha}{D}\Bigg\{&\Lambda_2G_2(\cosh\eta)+\Lambda_3\left[\frac{5G_4(\cosh\eta_\beta)}{G_1(\cosh\eta_\beta)}G_1(\cosh\eta)+G_4(\cosh\eta)\right]\\&+\Lambda_4H_2(\cosh\eta)\Bigg\}G_2(\cos\theta)\end{aligned} \tag{3.33}$$

where D, Λ_2, Λ_3 and Λ_4 are $\eta-$ and $\theta-$ dependent coefficients, defined by Dassios et al. [9], and $G_N(x)$ and $H_N(x)$ are the Gegenbauer polynomials of the first and second kind, respectively, of degree $-\frac{1}{2}$ and order N.

The Happel approach results in the algebraic linear system [9]:

$$\begin{bmatrix} G_1(\cosh\eta_a) & G_2(\cosh\eta_a) & G_4(\cosh\eta_a) & H_2(\cosh\eta_a)\\ G_1'(\cosh\eta_a) & G_2'(\cosh\eta_a) & G_4'(\cosh\eta_a) & H_2'(\cosh\eta_a)\\ G_1(\cosh\eta_\beta) & G_2(\cosh\eta_\beta) & G_4(\cosh\eta_\beta) & H_2(\cosh\eta_\beta)\\ KG_1''(\cosh\eta_\beta) & KG_2''(\cosh\eta_\beta) & KG_4''(\cosh\eta_\beta) & KH_2''(\cosh\eta_\beta)\\ +\Lambda G_1'(\cosh\eta_\beta) & +\Lambda G_2'(\cosh\eta_\beta) & +\Lambda G_4'(\cosh\eta_\beta) & +\Lambda H_2'(\cosh\eta_\beta)\end{bmatrix}\cdot \begin{bmatrix}A_1\\A_2\\A_3\\A_4\end{bmatrix}=\begin{bmatrix}MG_2(\cosh\eta_a)\\MG_1(\cosh\eta_a)\\0\\0\end{bmatrix} \tag{3.34}$$

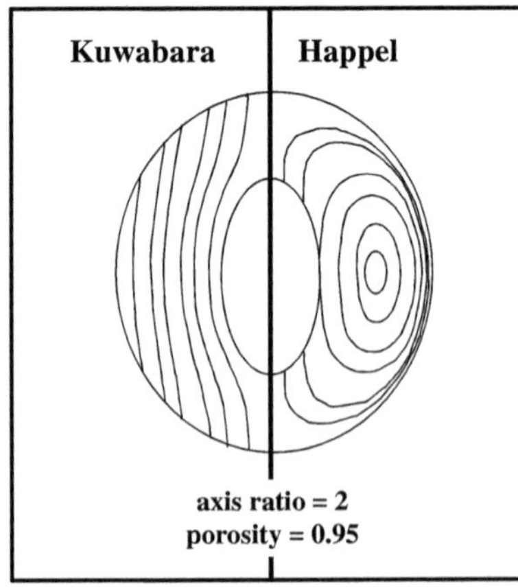

Fig. 3.5 Sample stream lines in two identical prolate spheroids-in-cell

where:

$$K = \cosh^2 \eta_\beta - \frac{1}{5} \tag{3.35}$$

$$\Lambda = -2 \cosh \eta_\beta \tag{3.36}$$

$$M = \frac{2}{\cosh^2 \eta_a - 1} \tag{3.37}$$

and where the primes and double primes denote the first and second derivatives of the corresponding functions, respectively.

Some typical results for the flow field in a spheroidal-in-cell are shown in Fig. 3.5.

The spheroid is oblate when $a_3 < a_1$, where the semifocal distance is given by $\bar{a} = \sqrt{a_1^2 - a_3^2}$. As shown by various researchers (see, for example, Happel and Brenner [12]) the stream function for the oblate case can be obtained from the same formulation as above, by using the following transformation from the prolate coordinates system (η, θ) to the oblate one $(\bar{\eta}, \theta)$:

$$\cosh \eta = i \cosh \bar{\eta} \tag{3.38}$$

$$a = -i\bar{a} \tag{3.39}$$

Following this, the solution for stream function can be easily obtained; some typical results of which are shown in Fig. 3.6.

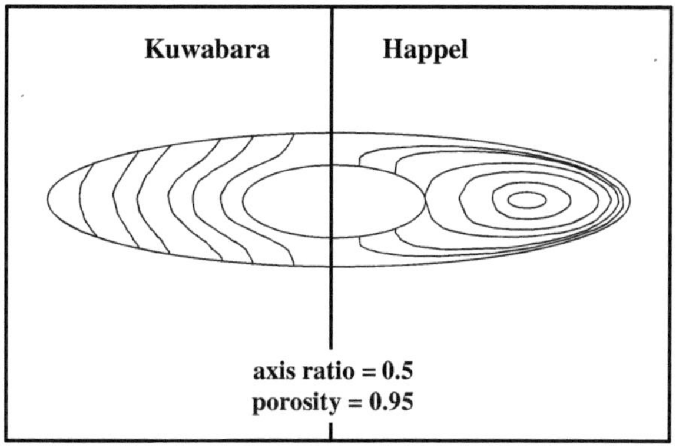

Fig. 3.6 Sample stream lines in two identical oblate spheroids-in-cell

3.3 Single Phase Flow in Granular Structures

In the presence of porous media, the need for a realistic description of the structure of a porous medium significantly increases the mathematical complexity of a model. As porous media are generally characterized by highly complex internal geometry, appropriate modeling is necessary for the derivation of meaningful conclusions. Numerous industrial and technological applications involving fluid flow and mass transport processes within multi-particle assemblages have attracted scientific interest in the last decades, mainly focusing on industrial physicochemical processes (sedimentation, catalysis, etc.), alternative energy sources (fuel cells, etc.), and separation techniques (chromatography, filters, etc.). Although arrays of regularly spatially distributed spheres represent an idealization of real granular media, they have been widely studied from both the fluid dynamics and mass transport points of view [19, 21]. On the other hand, due to their complex geometry, random particle distributions were the subject of rather limited investigations until the 1990s (see [2, 18]). Since then, fast advances in computational capabilities have contributed to reviving interest on this topic with emphasis placed on hydrodynamic aspects [5, 13, 22].

3.3.1 Representation of 3-D Sphere Assemblages

To define a realistic domain to solve flow and transport problems, a granular porous medium was constructed in the form of a spherical particle assemblage. The structure was digitized and the phase function (equal to zero for solid and unity for the pore space) was determined to obtain the specified porosity. More specifically, representation of the domains under consideration was achieved as follows:

Step (1) Using a random number generator, the position of the sphere's center was selected, being in a box of specified dimensions ($3 \times 2 \times 3$ mm)

Step (2) Using a random number generator, a radius assumed to follow the log-normal distribution was selected

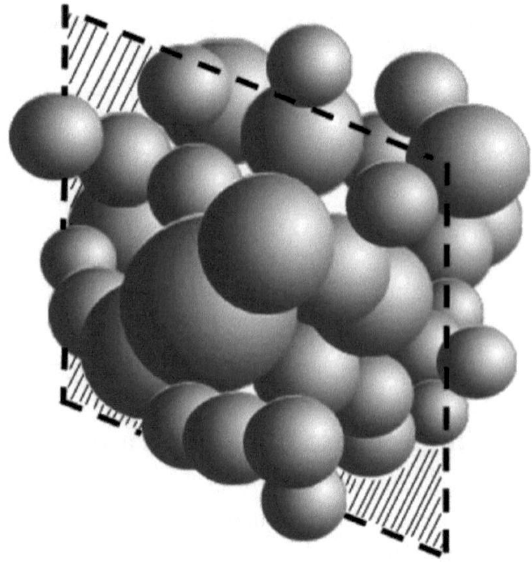

Fig. 3.7 A three-dimensional representation of a porous medium

Step (3) The void space around the sphere was checked. If it was free the radius value was accepted, otherwise Step 2 was repeated
Step (4) The sphere was posed
Step (5) Steps 1–4 were repeated until the volume of the positioned spheres satisfied the pre-defined porosity value.

Obviously, many three-dimensional representations can be generated by the above algorithm for a specific porosity value. Therefore, it is necessary to validate the results against these configurations to ensure that the macroscopic results are independent of each specific realization. A graphic representation of a representative medium for a typical porosity of $\varepsilon = 0.43$ is presented in Fig. 3.7.

A randomly selected two-dimensional cut of this domain in is shown in Fig. 3.8, which also depicts the grid of the numerical solution.

3.3.2 The Flow Field

For the numerical simulations, the velocity field was computed numerically by solving the Stokes equations

$$\nabla p = \mu \nabla^2 \mathbf{v} \tag{3.40}$$

$$\nabla \cdot \mathbf{v} = 0 \tag{3.41}$$

$$\mathbf{v} = 0 \text{ at liquid} - \text{solid interface} \tag{3.42}$$

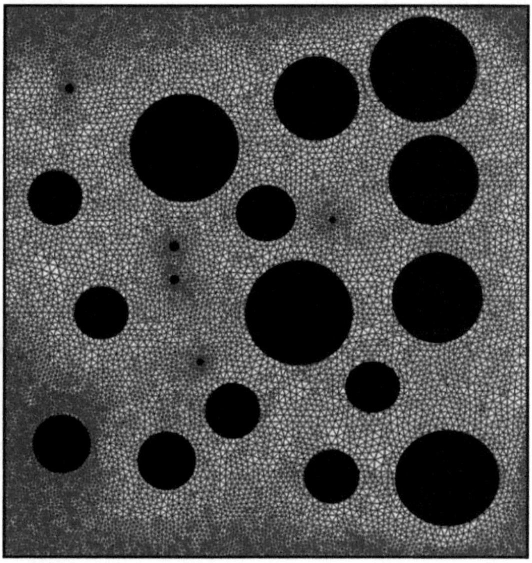

Fig. 3.8 A selected two-dimensional cut of the simulated geometry discretized by an unstructured grid

where $\mathbf{v}$, p, and μ are the velocity vector, pressure field, and fluid viscosity, respectively.

The procedure for solving the Stokes flow problem involves discretization in terms of cubic elements as follows [1, 6, 14, 15]: At the pore level, a staggered marker-and-cell (MAC) mesh is used, with the pressure defined at the center of the cell, and the velocity components defined along the corresponding face boundaries. The resulting linear system of equations is solved by a successive over-relaxation (SOR) method. An initial estimate of p is determined by solving the Laplace equation. Next, the velocity vector $\mathbf{v}$ is calculated from the corresponding momentum balance and the continuity equation $\nabla \cdot \mathbf{v} = 0$. The pressure is corrected through an artificial compressibility equation of the form:

$$\frac{dp}{dt} = \nabla \cdot \mathbf{v} \tag{3.43}$$

Essentially, the method adds an artificial density time derivative related to the pressure by an artificial equation of state $p = \beta\rho$, where β is an artificial compressibility factor. Similar to the compressible momentum equation, $c = \beta^{1/2}$ is an artificial speed of sound and for stability during the iterative procedure, its magnitude should be such that the respective artificial Mach number, $M = \frac{R}{c} \max_D \left(\sum_i u_i^2 \right)^{1/2}$ is low (M $\ll$ 1), where R is the relevant Reynolds number. In the limiting case of R $\to$ 0, which is the present case, any finite value of β should meet this criterion. Thus, $\beta = 1$ was selected although it is evident that the

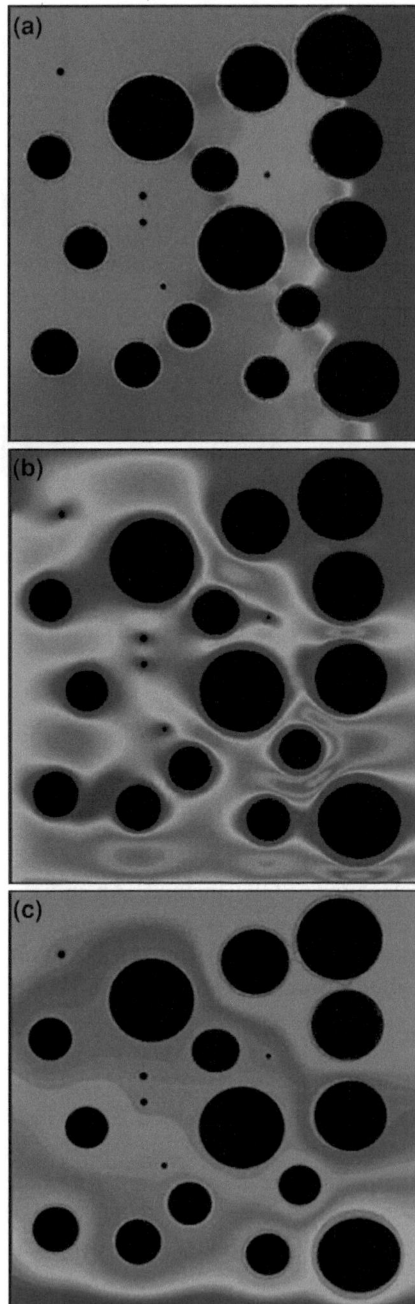

Fig. 3.9 Snapshot of the **a** pressure field, **b** velocity, and **c** stream line simulations through the representative porous medium. The flow direction is from *left* to *right*

exact value cannot have any effect on the final (steady state) results, since at steady state the artificial density time derivative is equal to zero.

The above steps are repeated until convergence is reached. This numerical scheme for the determination of the velocity field has been widely validated in terms of both the velocity field and the corresponding permeability [1, 6, 14, 15].

3.3.3 Results and Discussion

Figure 3.9 show the results of the (a) pressure field, (b) velocity, and (c) stream function for a typical porous medium of $\varepsilon = 0.72$. A randomly selected two-dimensional cut of the medium is considered so the results can be visualized clearly. The boundary condition at the closed walls is non-slip, at the left boundary an inflow was imposed, and at the right boundary an outflow condition. Small vortices and recirculating flow are produced in the medium depending on the pore size, while smoother profiles are obtained at the inlet and outlet. The velocity gradient observed from top to bottom at the inlet surface is generated because the inflow condition ensures constant molar flux instead of a plug-type velocity vector. Finally, it is interesting to observe that the faster flow paths are generated by the porous structure and they appear wherever pore diameters are fairly small.

References

1. Adler, P.M., Jacquin, C.J., Quiblier, J.A.: Flow in simulated porous media. Int J Multiphas Flow **16**, 691–712 (1990)
2. Berryman, J.G.: Effective conductivity by fluid analogy for a porous insulator filled with a conductor. Phys. Rev. B **27**, 7789–7792 (1983)
3. Brinkman, H.C.: A calculation of the viscous force exerted by a flowing fluid on a dense swarm of particles. Appl. Sci. Res. A **1**, 27–34 (1949)
4. Burganos, V.N., Coutelieris, F.A., Dassios, G., Payatakes, A.C.: On the rapid convergence of the analytical solution of stokes flow around spheroids-in-cells. Chem. Eng. Sci. **50**, 3313–3317 (1995)
5. Coelho, D., Thovert, J.F., Adler, P.M.: Geometrical and transport properties of random packings of spheres and aspherical particles. Phys. Rev. E **55**, 1959–1978 (1997)
6. Coutelieris, F.A., Kainourgiakis, M.E., Stubos, A.K.: Low peclet mass transport in assemblages of spherical particles for two different adsorption mechanisms. J. Colloid Interface Sci. **264**, 20–29 (2003)
7. Darcy H (1857) Recherches Experimentales Relatives au Mouvement de l'Eau dans les Tuyaux, Mallet-Bachelier, Paris
8. Dassios, G., Hadjinicolaou, M., Payatakes, A.C.: Generalized eigenfunctions and complete semiseparable solutions for stokes flow in spheroidal coordinates. Quart. Appl. Math. **52**, 157–191 (1994)
9. Dassios, G., Hadjinicolaou, M., Coutelieris, F.A., Payatakes, A.C.: Stokes flow in spheroidal particle-in-cell models with Happel and Kuwabara boundary conditions. Int. J. Eng. Sci. **33**, 1465–1490 (1995)

10. Happel, J.: Viscosity of suspensions of uniform spheres. J. Appl. Phys. **28**, 1288–1292 (1957)
11. Happel, J.: Viscous flow in multiparticle systems: Slow motion of fluids relative to beds of spherical particles. AIChE J. **4**, 197–201 (1958)
12. Happel, J., Brenner, H.: Low Reynolds Number Hydrodynamics. Prentice-Hall, Englewood Cliffs (1965)
13. Jullien, R., Meakin, P.: Simple three-dimensional models for ballistic deposition with restructuring. Europhys. Lett. **4**, 1385–1390 (1987)
14. Kainourgiakis, M.E., Kikkinides, E.S., Stubos, A.K.: Diffusion and flow in porous domains constructed using process-based and stochastic techniques. J. Porous. Mat. **9**, 141–154 (2002)
15. Kikkinides, E.S., Burganos, V.N.: Permeation properties of three-dimensional self-affine reconstructions of porous materials. Phys. Rev. E **62**, 6906–6915 (2000)
16. Klinkenberg, L.J.: The permeability of porous media to liquids and gases, drilling and production practice. Am. Petr. Inst. **V**, 200–213 (1941)
17. Kuwabara, S.: The forces experienced by randomly distributed parallel circular cylinder or spheres in a viscous flow at small Reynolds numbers. J. Phys. Soc. Jpn. **14**, 527–532 (1959)
18. Meakin, P., Skjeltorp, A.T.: Application of experimental and numerical models to the physics of multiparticle systems. Adv. Phys. **42**, 1–127 (1993)
19. Meredith, R.E., Tobias, C.W.: Resistance to potential flow through a cubical array of spheres. J. Appl. Phys. **31**, 1270–1273 (1960)
20. Moon, P., Spencer, D.E.: Field Theory Handbook: Including Coordinate Systems, Differential Equations, and Their Solutions, 2nd edn. Springer, Berlin (1971)
21. Rayleigh, R.S.: On the influence of obstacles arranged in rectangular order upon the properties of a medium. Philos. Mag. **34**, 481–502 (1892)
22. Visscher, W.M., Bolsterli, M.: Random packing of equal and unequal spheres in two and three dimensions. Nature **239**, 504–507 (1972)

Chapter 4
Transport Phenomena in Porous Structures

4.1 Introduction

The problem of solute dispersion during underground water movement has attracted interest from the early days of the last century [127], but it was only since the 1950s that the general topic of hydrodynamic dispersion, or miscible displacement, became the subject of more systematic study. This topic has interested hydrologists, geophysicists, petroleum and chemical engineers, among others, and for some time now it is treated at length in books on flow through porous media (e.g. [9, 116]). Some books on chemical reaction engineering (ex: [24, 55, 141]) treat the topic of dispersion (axial and lateral) in detail and it is generally observed that data for liquids and gases do not overlap, even in the "appropriate" dimensionless representation.

Since the early experiments of Slichter [127] and particularly since the analysis of dispersion during solute transport in capillary tubes, developed by Taylor [133] and Aris [5, 6], much work has been done on the description of the principles of solute transport in porous media of inert particles (ex: soils) and in packed bed reactors (see [9, 44]).

Gray [59], Bear [9] and Whitaker [143] derived the proper form of the transport equation for the average concentration of solute in a porous medium, by using the method of volume or spatial averaging, developed by Slattery [126].

Brenner [20] developed a general theory for determining the transport properties in spatially periodic porous media in the presence of convection, and showed that dispersion models are valid asymptotically in time for the case of dispersion in spatially periodic porous media, while Carbonell and Whitaker [26] demonstrated that this should be the case for any porous medium. These authors presented a volume–average approach for calculating the dispersion coefficient and carried out specific calculations for a two-dimensional spatially periodic porous medium. Eidsath et al. [48] have computed axial and lateral dispersion coefficients in packed beds based on these spatially periodic models, and have compared the

F. A. Coutelieris and J. M. P. Q. Delgado, *Transport Processes in Porous Media*, Advanced Structured Materials 20, DOI: 10.1007/978-3-642-27910-2_4,

results to available experimental data. The axial dispersion coefficient calculated by Eidsath et al. [48] shows a Peclet number dependence that is too strong, while their radial dispersion. However, in soils or underground reservoirs, large scale non-uniformities lead to values of dispersion coefficients that differ much from those measured in packed beds, and for these cases spatially periodic models cannot be expected to provide excellent results without modifications.

There have been other attempts at correlating and predicting dispersion coefficients based on a probabilistic approach [38, 60, 70, 114] where the network of pores in the porous medium is regarded as an array of cylindrical capillaries with parameters governed by probability distribution functions.

Dispersion in porous media has been studied by a significant number of investigators; using various experimental techniques. However, measurements of axial and lateral dispersion are normally carried out separately, and it is generally recognised that 'experiments on lateral dispersion are much more difficult to perform than those on axial dispersion' [116].

When a fluid is flowing through a bed of inert particles, one observes the dispersion of the fluid in consequence of the combined effects of molecular diffusion and convection in the spaces between particles. Generally, the dispersion coefficient in axial direction is superior to the dispersion coefficient in radial direction by a factor of 5, for values of Reynolds number larger than 10. For low values of the Reynolds number (say, $\mathrm{Re} < 1$), the two dispersion coefficients are approximately the same and e qual to molecular diffusion coefficient.

The detailed structure of a porous medium is greatly irregular and just some statistical properties are known. An exact solution to characterize flowing fluid through one of these structures is basically impossible. However, by the method of volume or spatial averaging it is possible to obtain the transport equation for the average concentration of solute in a porous medium [9, 143].

At a "macroscopic" level, the quantitative treatment of dispersion is currently based on Fick's law, with the appropriate dispersion coefficients; cross stream dispersion is related to the radial dispersion coefficient, D_{T}, whereas stream-wise dispersion is related to the axial dispersion coefficient, D_{L}.

If a small control volume is considered, a mass balance on the solute, without chemical reaction, leads to

$$D_{\mathrm{L}}\frac{\partial^2 C}{\partial z^2}+\frac{1}{r}\frac{\partial}{\partial r}\left(D_{\mathrm{T}}r\frac{\partial C}{\partial r}\right)-u\frac{\partial C}{\partial z}=\frac{\partial C}{\partial t} \tag{4.1}$$

where C is the mean solute concentration, $u\,(=U/\varepsilon$, where U is the superficial velocity and ε the porosity of the porous media of inert particles with diameter d) the mean interstitial velocity of fluid and t the time.

A large number of theories, namely the theories based on a probabilistic approach, have been proposed to explain dispersion in porous media; however, the theory of Saffman [114], who modelled the microstructure of a porous media as a network of capillary tubes of random orientation, and Koch and Brady [83] were the most referred.

4.2 Diffusion

Diffusion in porous media is a general subject that involves many fields of research, such as chemistry (e.g. porous catalytic pallets), biology (e.g. porous cellular organelles), and materials science (e.g. porous polymer matrixes for controlled-release and gas-storage materials).

Diffusivity or diffusion coefficient is a proportionality constant between the molar flux due to molecular diffusion and the gradient in the concentration of the species (or the driving force for diffusion). Diffusivity is encountered in Fick's law and numerous other equations of physical chemistry.

The effective diffusion coefficient, D_e, describes diffusion through the pore space of porous media. It is macroscopic in nature, because it is not individual pores but the entire pore space that needs to be considered. Under steady state conditions the mass flux, F, depends on the concentration gradient and is expressed by Fick's first law:

$$F = -D_e \frac{dC}{dx} \tag{4.2}$$

The effective diffusion coefficient for transport through the pores is estimated as follows:

$$D_e = \frac{D_m \varepsilon \delta}{\tau} \tag{4.3}$$

where τ and δ are dimensionless factors accounting for tortuosity (> 1) and constrictivity ($\leq$1) of the pores, respectively, and ε is the effective porosity which accounts for the reduced cross-sectional area available for diffusion when diffusion occurs only in the pore space. The constrictivity describes the slowing down of diffusion by increasing the viscosity in narrow pores as a result of greater proximity to the average pore wall. It is a function of pore diameter and the size of the diffusing particles.

4.2.1 Constrictivity Factor

The constrictivity factor δ depends on the ratio of the solute diameter to the pore diameter (λ_p):

$$\lambda_p = \frac{\text{molecule diameter}}{\text{pore diameter}} \tag{4.4}$$

may be quantified by empirical equations, as developed by Beck and Schultz [11] and Chantong and Massoth [28], respectively:

$$\delta = \left(1 - \lambda_p\right)^4 \tag{4.5}$$

$$\delta = 1.03\exp(-4.05\lambda_p) \tag{4.6}$$

4.2.2 Tortuosity Factor

The tortuosity factor, τ, in Eq. 4.3 accounts for the pore geometry and is defined as the square of the ratio of the effective path length, le, in the pore to the shortest distance, l, in a porous medium:

$$\tau = \left(\frac{le}{l}\right)^2 \tag{4.7}$$

Tortuosity is not a physical constant and depends first of all on other porous media characteristics, like porosity, pore diameter, channel shape, etc. In general, in granular packings or beds the value of tortuosity lies in the region 1.1–1.7 (see [9] and Dullien [43]), and our experimental results obtained with packed beds of narrow size particles are in this range.

Several empirical correlations, suggesting a relationship between tortuosity and porosity, have been found in the literature since the end of the nineteen century:

$$\tau = 1.5 - 0.5\varepsilon \quad \text{Ref. [113]} \tag{4.8}$$

$$\tau = \varepsilon^{-n} \quad \text{Ref. [4]} \tag{4.9}$$

$$\tau = 1 - 0.5 \times \ln(\varepsilon) \quad \text{Ref. [168]} \tag{4.10}$$

$$\tau = \sqrt{1 - \ln(\varepsilon^2)} \quad \text{Ref. [22]} \tag{4.11}$$

$$\tau = 1 - 0.41 \times \ln(\varepsilon) \quad \text{Ref. [42]} \tag{4.12}$$

$$\tau = 1/\sqrt{\varepsilon} \quad \text{Refs. [27], [113]} \tag{4.13}$$

Recently, Yun et al. [148] presented a theoretical geometry model for tortuosity of tortuous streamtubes in a porous media with spherical particles. The authors suggested an average value of tortuosity given by

$$\tau = \frac{\tau_1 + \tau_2}{2} \tag{4.14}$$

with

$$\tau_1 = \frac{\left[1 + \frac{\sqrt{3}(\pi-2)}{6+3P}\right] + \left[\frac{\sqrt{2+3P^2/4+3P} + \arcsin\left(\sqrt{3}(2+P)/2\right)^{-1}}{\sqrt{3}(2+P)/2}\right]}{4} + \frac{\frac{1}{\sqrt{1-\left[\frac{1}{\sqrt{3}(2+P)/2}\right]^2}} + \frac{1}{\sqrt{1-\left[\frac{1}{\sqrt{3}(2+P)/2-1}\right]^2}}}{4} \tag{4.15}$$

$$\tau_2 = \left(1 - \frac{P_1}{P_1+1}\right)\left(1 + \frac{\pi - 2}{P_1+2}\right) + \frac{P_1}{P_1+1} \tag{4.16}$$

where $P = \sqrt{2\pi/\left(\sqrt{3}(1-\varepsilon)\right)} - 2$ and $P_1 = \sqrt{\pi/(1-\varepsilon)} - 2$.

Equations 4.8–4.14 all satisfy the condition $\tau = 1$ for $\varepsilon = 1$, and this is consistent with the physical situation observed. Yun et al. [148] and Sen et al. [120] showed that for an isotropic medium with spherical particles the tortuosity of porous and granular media decreases with increasing bed voidage and increase for non-spherical particles.

4.2.3 *Porosity Factor*

The overall porosity, ε, of porous media can be determined (pore size distribution and tortuosities are unknown). Therefore the relative diffusivity ($D^* = D_e/D_m$) is often defined as an empirical function of ε alone ($\delta = 1$):

$$D_e = D_m \varepsilon^m \tag{4.17}$$

where m is an empirical exponent.

Archie [4] found that the exponent m varied between 1.8 and 2.0 in consolidated materials. In unconsolidated sand he found a value of 1.3. For an isotropic packing of spherical particles a theoretical value of m = 3/2 was derived [21]. Adler et al. [1] reported a value of m = 1.64 for Fontainebleau sandstone. Probst and Wohlfahrt [107] found that m equals 1.43 for loose packings of catalyst particles and Millington and Quirk [96] reported a value of 4/3 for diffusive flow of gases at normal pressures or diffusion of ions in solution in soils. In experiments on the diffusion of gases through compacted sands a value of m = 1.5 was determined [122].

4.2.4 *Diffusion in Semi-Infinite Porous Media*

In the limit of very low fluid velocity, where $u \to 0$, dispersion is determined solely by molecular diffusion, with $D_T = D_L = D_m/\tau = D_e$. Diffusion may be the dominant mass transfer mechanism (compared to advection) in zones of low hydraulic conductivity. The geometry of these layers may be considered as plane sheets and analytical solutions of Fick's second law are available for a variety of initial and boundary conditions [36].

Thick confining layers and bedrock formations of low permeability may represent a sink for contaminants which are spread out in an aquifer or a landfill. On the other hand, once such low-permeability domains are contaminated, they may become a long-term source during remediation of the aquifer (e.g. pump-and-treat). These formations can be considered as semi-infinite media for diffusion. If the low conductivity zone is free of the contaminant initially and then exposed to a constant concentration at the surface for a given period of time, the initial and boundary conditions are:

$$t = 0 \quad x > 0 \quad C = C_1 \tag{4.18a}$$

$$t > 0 \quad x = 0 \quad C = C_0 \tag{4.18b}$$

$$t > 0 \quad x \to \infty \quad C = C_1 \tag{4.18c}$$

The concentration profile at a given time is:

$$\frac{C - C_1}{C_0 - C_1} = 1 - erf\left(\frac{x}{2\sqrt{D_e t}}\right) \tag{4.19}$$

The quantity $2\sqrt{D_e t} = d_p$ can be considered as "penetration depth" which represents the time dependent distance within which 87% of the mass of the diffusing substance occurs. Accordingly, the "penetration time" then represents the time after which 87% of the diffusing molecules have not yet moved beyond a given distance.

The special case of zero surface concentration is obvious. The rate of loss diffusing substance from the semi-infinite medium when the surface concentration is zero, is

$$D_e\left(\frac{\partial C}{\partial x}\right)_{x=0} = \frac{D_e C_0}{\sqrt{D_e t}} \tag{4.20}$$

and the total amount M_t of diffusing substance which has left the medium at time t is given by integrating Eq. 4.19 and is

$$M_t = 2C_0\sqrt{\frac{D_e t}{\pi}} \tag{4.21}$$

4.2.5 Diffusion in a Plane Sheet

In this case we consider one-dimensional diffusion through a plane sheet of thickness $2d$, initially at a uniform concentration C_0 and with the surface at constant concentration C_1.

The boundary conditions are:

$$t = 0 \quad -d < x < d \quad C = C_1 \tag{4.22a}$$

$$t > 0 \quad x = 0 \quad C = C_0 \tag{4.22b}$$

$$t > 0 \quad x \to \infty \quad C = C_1 \tag{4.22c}$$

The mass of solute per unit area which has diffused after a certain time is:

$$\frac{C - C_1}{C_0 - C_1} = 1 - \frac{4}{\pi}\sum_{n=0}^{\infty}\frac{(-1)^n}{2n+1}\cos\left(\frac{(2n+1)\pi x}{2d}\right)\exp\left(\frac{(2n+1)^2\pi^2 D_e t}{4d^2}\right) \tag{4.23}$$

and the total amount M_t of diffusing substance which has entered the plane sheet at time t is given by

$$\frac{M_t}{M\infty} = 1 - \sum_{n=0}^{\infty}\frac{8}{(2n+1)^2\pi^2}\exp\left(\frac{(2n+1)^2\pi^2 D_e t}{4d^2}\right) \tag{4.24}$$

where M_∞ is the corresponding quantity after infinite time. The corresponding solutions for small times are:

$$\frac{C - C_1}{C_0 - C_1} = \sum_{n=0}^{\infty}(-1)^n erfc\frac{(2n+1)d - x}{2\sqrt{D_e t}} + \sum_{n=0}^{\infty}(-1)^n erfc\frac{(2n+1)d + x}{2\sqrt{D_e t}} \tag{4.25}$$

and

$$\frac{M_t}{M\infty} = \frac{2}{d}\sqrt{D_e t}\left[\pi^{-1/2} + 2\sum_{n=1}^{\infty}(-1)^n ierfc\frac{nd}{\sqrt{D_e t}}\right] \tag{4.26}$$

4.2.6 Diffusion in a Cylinder

If we consider a circular cylinder in which the diffusion is radial, concentration is then a function of radius and time, and the diffusion equation is obtained:

$$\frac{\partial C}{\partial t} = \frac{1}{r}\frac{\partial}{\partial r}\left(rD_e\frac{\partial C}{\partial r}\right) \tag{4.27}$$

Considering the following boundary conditions:

$$t = 0 \quad 0 < r < a \quad C = C_1 \tag{4.28a}$$

$$t > 0 \quad r = a \quad C = C_0 \tag{4.28b}$$

$$t > 0 \quad r \to \infty \quad C = C_1 \tag{4.28c}$$

the solution obtained is:

$$\frac{C - C_1}{C_0 - C_1} = 1 - \frac{2}{a}\sum_{n=1}^{\infty} \frac{1}{\alpha_n J_1(a\alpha_n)} \exp(-D_e \alpha_n^2 t) J_0(r\alpha_n) \tag{4.29}$$

where α_n is the positive roots of $J_0(a\alpha_n) = 0$, $J_0(x)$ is the Bessel function of the first kind of order zero, and $J_1(x)$ is the Bessel function of the first order.

The total amount M_t of diffusing substance which has entered or left the cylinder at time t is given by

$$\frac{M_t}{M\infty} = 1 - \sum_{n=1}^{\infty} \frac{4}{a^2\alpha_n^2} \exp(-D_e \alpha_n^2 t) \tag{4.30}$$

where M_∞ is the corresponding quantity after infinite time. The corresponding solutions for small times are:

$$\begin{aligned} \frac{C - C_1}{C_0 - C_1} = {} & \sqrt{\frac{a}{r}}\, erfc \frac{a - r}{2\sqrt{D_e t}} + \frac{(a - r)\sqrt{D_e a t}}{4ar^{3/2}}\, ierfc \frac{a - r}{2\sqrt{D_e t}} \\ & + \frac{(9a^2 - 7r^2 - 2ar)D_e t}{32a^{3/2}r^{5/2}}\, i^2 erfc \frac{a - r}{2\sqrt{D_e t}} + \ldots \end{aligned} \tag{4.31}$$

and

$$\frac{M_t}{M\infty} = \frac{4}{\sqrt{\pi}}\sqrt{\frac{D_e}{a^2}t} - \frac{D_e}{a^2}t - \frac{1}{3\sqrt{\pi}}\left(\frac{D_e}{a^2}t\right)^{3/2} + \ldots \tag{4.32}$$

4.2.7 Diffusion in a Sphere

If we consider a sphere in which the diffusion is radial, concentration is then a function of radius and time, and the diffusion equation is obtained:

$$\frac{\partial C}{\partial t} = D_e \frac{\partial^2 C}{\partial r^2} + \frac{2}{r} D_e \frac{\partial C}{\partial r} \tag{4.33}$$

Considering the following boundary conditions:

$$t = 0 \quad 0 < r < a \quad C = C_1 \tag{4.34a}$$

$$t > 0 \quad r = a \quad C = C_0 \tag{4.34b}$$

$$t > 0 \quad r \to \infty \quad C = C_1 \tag{4.34c}$$

the solution obtained is:

$$\frac{C - C_1}{C_0 - C_1} = 1 + \frac{2a}{\pi r}\sum_{n=1}^{\infty}\frac{(-1)^n}{n}\sin\frac{n\pi r}{a}\exp\left(-\frac{D_e n^2\pi^2}{a^2}t\right) \tag{4.35}$$

And the concentration at the centre of the sphere, $r \to 0$, is given by

$$\frac{C - C_1}{C_0 - C_1} = 1 + 2\sum_{n=1}^{\infty}(-1)^n\exp\left(-\frac{D_e n^2\pi^2}{a^2}t\right) \tag{4.36}$$

The total amount M_t of diffusing substance which has entered or left the sphere at time t is given by

$$\frac{M_t}{M\infty} = 1 - \frac{6}{\pi^2}\sum_{n=1}^{\infty}\frac{1}{n^2}\exp\left(-\frac{D_e n^2\pi^2}{a^2}t\right) \tag{4.37}$$

where M_∞ is the corresponding quantity after infinite time. The corresponding solutions for small times are:

$$\frac{C - C_1}{C_0 - C_1} = \frac{a}{r}\sum_{n=0}^{\infty}\left[erfc\frac{(2n+1)a - r}{2\sqrt{D_e t}} - erfc\frac{(2n+1)a + r}{2\sqrt{D_e t}}\right] \tag{4.38}$$

and

$$\frac{M_t}{M\infty} = 6\sqrt{\frac{D_e}{a^2}t}\left[\pi^{-1/2} + 2\sum_{n=0}^{\infty} ierfc\frac{na}{\sqrt{D_e t}}\right] - 3\frac{D_e}{a^2}t \tag{4.39}$$

4.3 Axial Dispersion

Over the past 5 decades, axial dispersion in porous media has been measured and correlated extensively for gaseous and liquid systems. Many publications are available for a variety of applications, including: packed bed reactors [29, 47, 65, 89, 135] and soil column systems [9, 99, 102, 103].

One of the first results published about axial dispersion in packed beds of inertial particles was in the 1950s by Danckwerts [37], who published his celebrated paper on residence time distribution in continuous contacting vessels, including chemical reactors, and thus provided methods for measuring axial dispersion rates. The author studied dispersion along the direction of flow for a step input in solute concentration (C_S) in a bed of Raschig rings (with length L), crossed by water (C_0) with a value of $\mathrm{Re}(= \rho U d/\mu)$ approximately equal to 25 and obtained a $\mathrm{Pe_L}(= ud/D_L)$ value of 0.52.

Kramers and Alberda [85] followed Danckwerts's study with a theoretical and experimental investigation by the response to a sinusoidal input signal. These authors proposed that packed beds could be represented as consecutive regions of well-mixing rather than a sequence of stirred tanks (mixing-cell model) and suggested a $\mathrm{Pe_L} \cong 1$, for $\mathrm{Re} \to \infty$. McHenry and Wilhelm [94] assumed the axial distance between the mixing-cells in a packing to be equal to particle diameter and showed that Pe_L must be about two for high Reynolds number. The difference in the two results may be explained on the basis of experimental results of Kramers and Alberda [85] while are obtained with $L/D \approx 4.6$, a value significantly less than $L/D > 20$ [66]. Klinkenberg et al. [81] and Bruinzeel et al. [22] show that radial dispersion can be neglected for a small ratio of column diameter to length and large fluid velocity.

Brenner [19] presented the solution of a mathematical model of dispersion for a bed with finite length, L, and the most relevant conclusion of his work was that for $Pe_a(= uL/D_L) \geq 10$, the equations obtained by Danckwerts [37] for an input step in solute concentration and Levenspiel and Smith [89] for a pulse in solute concentration, that assumed an infinite bed, are corrected.

Hiby [75] proposed a better empirical correlation to cover the range of Reynolds numbers to 100. The author reported experimental results with the aid of photographs to compare the two dispersion mechanisms presented above: diffusional model in turbulent flow and the mixing-cell model.

Sinclair and Potter [124] used a frequency response technique applied to the flow of air through beds of glass ballotini in a Reynolds number range between 0.1 and 20. A further investigation in the intermediate Reynolds number region has been carried out by Evans and Kenney [50] who used a pulse response technique in beds of glass spheres and Raschig rings.

Experiments reported by Gunn and Pryce [68] showed that axial dispersion coefficients given by the theoretical equation for the diffusional model and the theoretical equation for the mixing-cell model are very similar. The authors also showed that neither the mixing-cell model nor the axially dispersed plug flow model could describe axial dispersion phenomena.

The description of solute transport in packed beds by dispersion models has been studied since the 1950s and has long attracted the attention of engineers and scientists.

Typically, the boundary conditions adopted, by the vast majority of the investigators reported above, have corresponded to the semi-infinite bed, i.e., L is sufficiently large ($L/D > 20$). Dispersion of the given tracer was measured at two points in the outlet and the distortion of a tracer forced by a pulse input (ex: [13, 25, 128]), frequency response (ex: [39, 45, 85, 94, 130]) and step input (ex: [37, 75, 95, 100]). Figure 4.1 illustrates some experimental data points for axial dispersion in liquid and gaseous systems. The experimental dispersion data are most frequently presented in logarithmic plots of Pe_L (or D_L/D_m) *vs.* $Pe_m(= ud/D_m)$, spanning six or more orders of magnitude.

4.3.1 Parameters Influencing Axial Dispersion: Porous Medium

Perkins and Johnston [102] in their article review showed some of the variables that influence axial and radial dispersion. However, before attempting in the

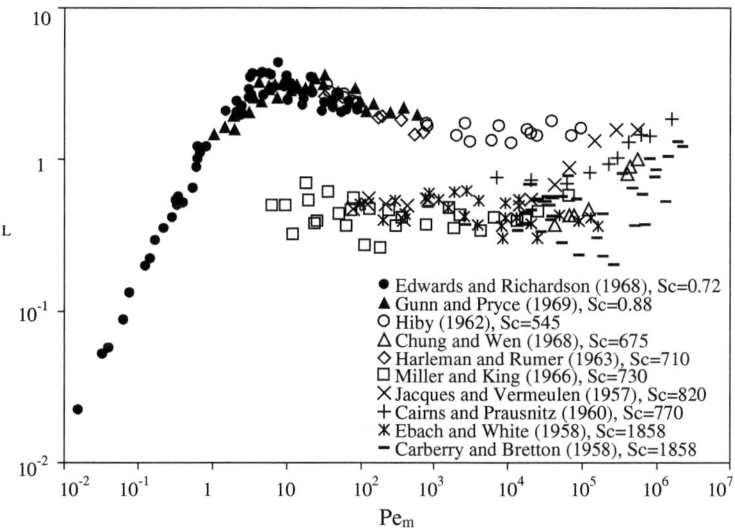

Fig. 4.1 Some experimental data points for axial dispersion in liquid systems and gaseous systems

parameters influencing dispersion, it is important to consider the effect of the packing of the bed on dispersion coefficients. Gunn and Pryce [68] and Roemer et al. [111] showed that when particles in packed beds are not well packed the dispersion coefficient is increased. Experimental results of Gunn and Pryce [68] showed that different re-packing of the bed gave deviations of 15% in radial Peclet values. These experiments confirm that fluid mechanical characteristics are not only defined by the values of the porosity and tortuosity (easy to reproduce), but depend of the quality of packing in the bed.

The effect of radial variations of porosity and velocity on axial and radial transport of mass in packed beds was analytically quantified by Choudhary et al. [30], Lerou and Froment [88], Vortmeyer and Winter [138] and Delmas and Froment [41].

A rigorous measurement of the porosity in a packed bed is fundamental to minimize the errors in the experimental measurements, because the porosity between the inert particles of the bed helps the diffusion of a tracer and gradually increases dispersion.

A more coherent interpretation of the experimental data may be obtained through the use of dimensional analysis. As a starting point it is reasonable to accept the functional dependence

$$D_L = \phi(L, D, u, d, \rho, \mu, D_m) \tag{4.40}$$

for randomly packed beds of mono sized particles with diameter d, where ρ and μ are the density and viscosity of the liquid, respectively, and D_m is the coefficient of molecular diffusion of the solute. Making use of Buckingham's π theorem, Eq. 4.40 may be rearranged to give

$$\frac{D_{\mathrm{L}}}{D_{\mathrm{m}}} \text{ or } \mathrm{Pe_L} = \Phi\left(\frac{L}{D}, \frac{D}{d}, \frac{ud}{D_{\mathrm{m}}}, \frac{\mu}{\rho D_{\mathrm{m}}}\right) \tag{4.41}$$

and it is useful to define $\mathrm{Pe_m} = ud/D_{\mathrm{m}}$ and $\mathrm{Sc} = \mu/\rho D_{\mathrm{m}}$. This result suggests that experimental data be plotted as $(D_{\mathrm{L}}/D_{\mathrm{m}})$ *vs.* $\mathrm{Pe_m}$.

4.3.1.1 Effect of Column Length

One first aspect to be considered, as a check on the experimental method (infinite medium), is the influence of the length of the bed (L) on the measured value of axial dispersion. In reality, if an experimental method is valid, values of the dispersion coefficient measured with different column lengths, under otherwise similar conditions, should be equal, within the reproducibility limits.

The dependence of the axial dispersion coefficient on the position in packed beds was first examined by Taylor [133]. The author showed that, in laminar flow, dispersion approximation would be valid if the following equation is satisfied,

$$\theta = \frac{D_{\mathrm{m}} t}{R^2} \gg 0.14 \tag{4.42}$$

where R is the tube radius. Carbonell and Whitaker [26] concluded that the axial dispersion coefficient becomes constant if the following expression is satisfied

$$\theta = \left(\frac{1-\varepsilon}{\varepsilon}\right)^2 \frac{D_{\mathrm{m}} t}{d^2} \gg 1 \tag{4.43}$$

Han et al. [69], see Fig. 4.2, showed that values of the axial dispersion coefficient, for uniform size packed beds, measured at different positions in the bed are function of bed location unless the approximate criterion

$$\frac{L}{d}\frac{1}{\mathrm{Pe_m}}\left(\frac{1-\varepsilon}{\varepsilon}\right)^2 \geq 0.3 \quad \text{or} \quad \theta = \frac{D_{\mathrm{m}} t}{d^2} \geq 0.15 \tag{4.44}$$

is satisfied. The authors showed that for $\mathrm{Pe_m} < 700$, axial dispersion coefficients were nearly identical for all values of $x = L$, and for $\mathrm{Pe_m} > 700$ observed an increase in the value of dispersion coefficients with increasing distance down the column.

4.3.1.2 Ratio of Column Diameter to Particle Diameter

It is well known (e.g. [137]) that the voidage of a packed bed (and therefore the fluid velocity) is higher near a containing flat wall. The effects of radial variations

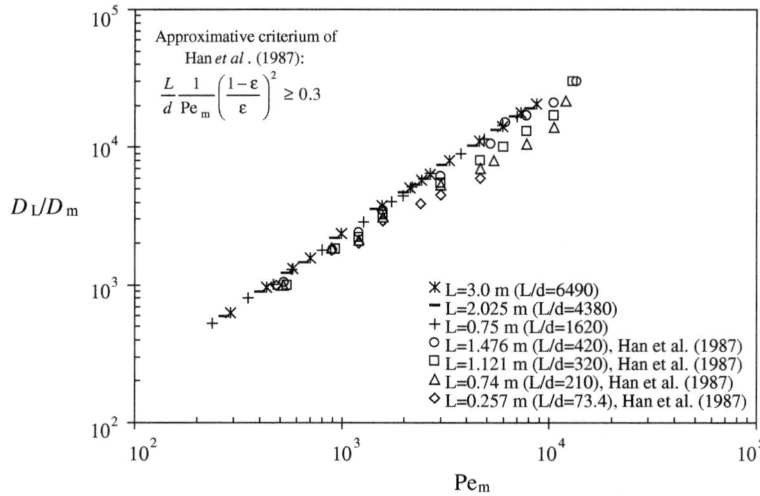

Fig. 4.2 Effect of bed length on axial dispersion

of porosity and velocity on axial and radial transport of mass in packed beds were analytically quantified by several investigators like Choudhary et al. [30], Lerou and Froment [88], Vortmeyer and Winter [138] and Delmas and Froment [41].

Schwartz and Smith [118] were the first to present experimental data showing zones of high porosity extending two or three particle diameters from the containing flat wall. The results indicated that unless $D/d > 30$ important velocity variations exist across the packed bed. Other studies showed that packed bed velocity profiles significantly differ from flows with large diameter particles in small diameter tubes ([23, 34]).

Hiby [75] showed that the effect of D/d is not significant in the measured of axial dispersion coefficient when the ratio is greater than 12.

Stephenson and Stewart [129] showed that the area of high fluid velocities limits to the area of high porosities, and this area does not extend more than a particle diameter of the wall and the assumption of a flat velocity profile is reasonable. This work confirms the earlier experiments reported by Roblee et al. [110], Schuster and Vortmeyer [117] and Vortmeyer and Schuster [137].

A similar effect was observed in measuring pressure drops across packings, so an empirical rule can be considered that the variations, in radial position, of the fluid velocity, porosity and dispersion coefficient can be negligible, if $D/d > 15$ [3, 66].

4.3.1.3 Ratio of Column Length to Particle Diameter

Strang and Geankoplis [130] and Liles and Geankoplis [90] make much of the effect of L/d but the evidence from fluid mechanical studies [67] was that the effect is confined to a dozen layers of particles and is not very important.

Experimental results of Guedes de Carvalho and Delgado [61], presented in Fig. 4.3, with two different spherical particles diameter and the same length of the packed bed showed that axial dispersion coefficient does not increased with particle diameter, as long as the condition $D/d > 15$ is satisfied (see Vortmeyer and Schuster [137] and Ahn et al. [2], for wall effects).

4.3.1.4 Particle Size Distribution

Another aspect of dispersion in packed beds that needs to receive attention is the effect of porous medium structure. In a packed bed of different particle sizes, the small particles accumulate in the interstices between large particles, and porosity tends to decrease.

Raimondi et al. [108] and Niemann [98] studied the effect of particle size distribution on axial dispersion and concluded that D_{L} increases with a wide particle size distribution. Eidsath et al. [48] indicated a strong effect of particle size distribution on dispersion. As the ratio of particle diameters went from a value of 2 to 5, the axial dispersion increased by a factor of 1.5, and radial dispersion decreased by about the same factor.

Han et al. [69] showed that for a size distribution with a ratio of maximum to minimum particle diameter equal to 7.3, axial dispersion coefficient are 2–3 times larger than the uniform size particles (see Fig. 4.3).

Wronski and Molga [146] studied the effect of particle size non-uniformities on axial dispersion coefficients during laminar liquid flow through packed beds (with a ratio of maximum to minimum particle diameter equal to 2.13) and proposed a generalized function to determine the increase of the axial dispersion coefficients in non-uniform beds relative to those obtained in uniform beds.

Guedes de Carvalho and Delgado [61] obtained the same conclusion in their experiments, with ballotini and a ratio of maximum to minimum particle diameter equal to 3.5 in comparison with glass ballotini that have the same size.

4.3.1.5 Particle Shape

The effect of particle shape on axial dispersion has been studied by several investigators, such as Bernard and Wilhelm [14], Ebach and White [45], Carberry and Bretton [25], Strang and Geankopolis [130], Hiby [75], Klotz [82] and more recently Guedes de Carvalho and Delgado [61]. The authors have used beds of spheres, cubes, Raschig rings, sand, saddles and other granular material, and have concluded that generally axial dispersion coefficient tend to be greater with packs of nonspherical particles than with packs of spherical particles, with the same size.

Figure 4.4 shows that particle shape is a significant parameter, with higher values of D_{L} (i.e., lower $\mathrm{Pe_L}$) being observed in packed beds of sand and Raschig rings comparatively with the results obtained with spherical beds.

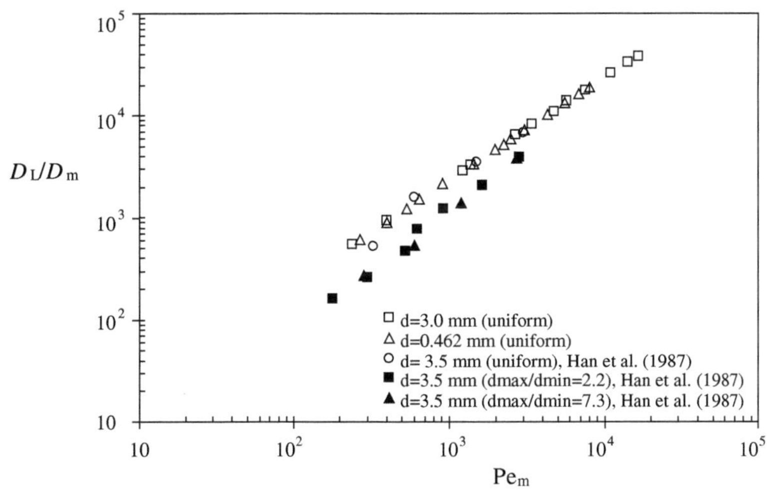

Fig. 4.3 Effect of particle size distribution on axial dispersion

Therefore, increased particle sphericity correlates with decreased dispersion, with a sphericity defined as the surface area of a particle divided by the surface area of a sphere of volume equal to the particle.

4.3.2 Parameters Influencing Axial Dispersion: Fluid Properties

4.3.2.1 Viscosity and Density of the Fluid

Some investigators, like Hennico et al. [74], used glycerol and obtained significant effect of viscosity, at large Reynolds number, on axial dispersion coefficient. In vertical miscible displacements, if a less viscous fluid displaced another fluid viscous fingers will be formed [102]. However, if a more viscous fluid displaced a different fluid the dispersion mechanisms are unaffected, but the situation will tend to reduce convective dispersion. This leads to increased dispersion relative to the more viscous fluid displacing a less viscous one.

The importance of density gradients was recently investigated by Benneker et al. [12] and their experiments showed that axial dispersion coefficient is considerably affected by fluids with different densities due the action of gravity forces. Fluid density creates similar effects to fluid viscosity. In a displacement with a denser fluid above the less-dense fluid, gravity forces cause redistribution of the fluids. However, if a denser fluid is on the bottom, usually, a stable displacement occurs.

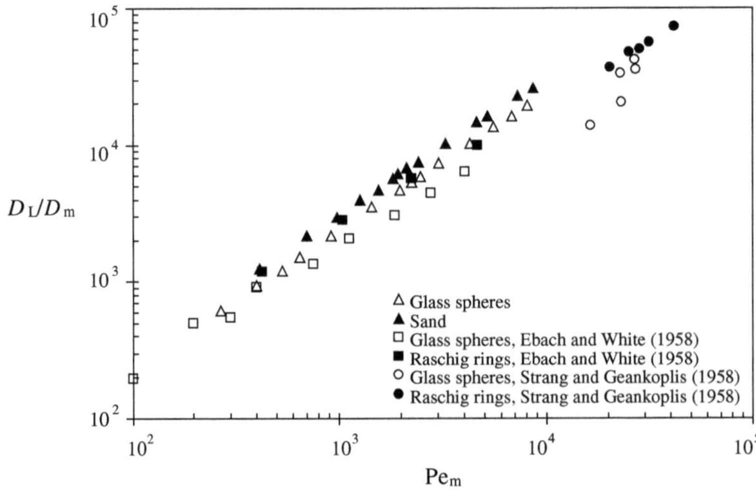

Fig. 4.4 Effect of particle shape on axial dispersion

4.3.2.2 Fluid Velocity

The first two groups of Eq. 4.8 have importance only when D/d is less than 15 and L/D is so small that the characteristics of dispersion are affected by changing velocity distributions. So, for packed beds we will usually have $D_L/D_m = \Phi(\mathrm{Pe_m}, \mathrm{Sc})$.

In order to understand the influence of fluid velocity on the dispersion coefficient, it is important to consider the limiting case where $u \to 0$. If D_L was defined based on the area open to diffusion (see Eq. 2), in the limit $u \to 0$, solute dispersion is determined by molecular diffusion, with $D_L = D_m' = D_m/\tau$ (τ being the tortuosity factor for diffusion and it is equal to $\sqrt{2}$ as suggested by Sherwood et al. [121]).

As the velocity of the fluid is increased, the contribution of convective dispersion becomes dominant over that of molecular diffusion (see [144]) and $D_L = ud/\mathrm{Pe_L}(\infty)$, where u is the interstitial fluid velocity and $\mathrm{Pe_L}(\infty) \cong 2$ for gas or liquid flow through beds of (approximately) isometric particles, with diameter d [23, 79].

Assuming that the diffusive and convective components of dispersion are additive, the same authors suggest that $D_L = D_m' + ud/\mathrm{Pe_L}(\infty)$, which may be written in dimensionless form [66] as

$$\frac{D_L}{D_m} = \frac{1}{\tau} + \frac{1}{2}\frac{ud}{D_m} \quad \text{or} \quad \frac{1}{\mathrm{Pe_L}} = \frac{1}{\tau}\frac{\varepsilon}{\mathrm{ReSc}} + \frac{1}{2} \tag{4.45}$$

This equation is expected to give the correct asymptotic behaviour in gas and liquid flow through packed beds, at high and low values of $\mathrm{Pe_m}(=ud/D_m)$. For gases this is confirmed in Fig. 4.5, but for liquids (Fig. 4.6) the data do not cover the extreme conditions.

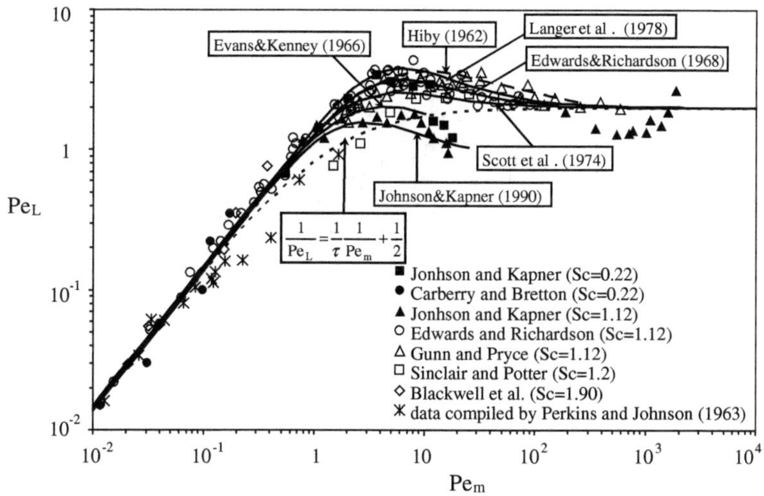

Fig. 4.5 Axial dispersion in gases

But these figures show that Eq. 4.45 is inaccurate over part of the intermediate range of Pe_m. In the case of gas flow, shown Fig. 4.5, significant deviations are observed only in the range $0.6 < Pe_m < 60$, as pointed out by several of the authors [47, 66, 75, 136]. The experimental values of $Pe_L (= ud/D_L)$ are generally higher than predicted by Eq. 4.12. Several equations have been proposed to represent the data in this intermediate range and the equations of Hiby [75], Edwards and Richardson [47], Evans and Kenney [50], Scott et al. [119], Langer et al. [80] and Johnson and Kapner [80] are shown to fit the data points reasonably well (see Fig. 4.5).

With liquids, deviations from Eq. 4.45 occur over the much wider range $2 < Pe_m < 10^6$, the experimental values of Pe_L being significantly lower than predicted by that equation. The difference in behaviour between gases and liquids has to be ascribed to the dependence of Pe_L on Sc $(= \mu/\rho D_m)$.

4.3.2.3 Fluid Temperature (or Schmidt Number)

The coefficient of axial dispersion for gas flow (Sc $\cong$ 1) is predicted with good accuracy by Eq. 4.45, except in the approximate range $0.5 < Pe_m < 100$, where experimental values may be more than twice those given by the equation, as confirmed by Fig. 4.5.

For liquid flow, a large number of data are available, that were obtained with different solutes in water at near ambient temperature, corresponding to values of Sc in the range $500 < Sc < 2000$. Most of the data reported in the literature, for this range of Sc, are shown in Fig. 4.6, and they form a "thick cloud" running parallel

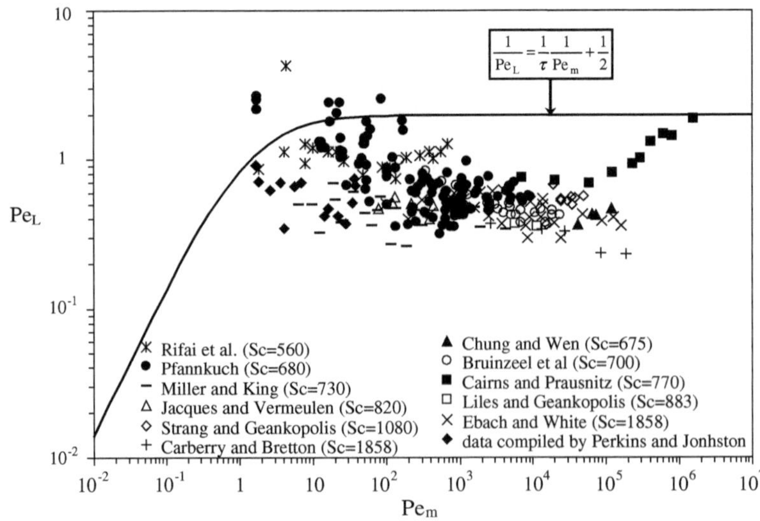

Fig. 4.6 Axial dispersion in liquids

to the line defined by Eq. 4.45, though somewhat below it (at approximately, $0.3 < Pe_L < 2$).

In recent years, data on axial dispersion have been made available for values of Sc between the two extremes of near ideal gas ($Sc \cong 1$) and cold water ($Sc > 550$). Such data were obtained either supercritical carbon dioxide ($1.5 < Sc < 20$) or heated water ($55 < Sc < 550$) and are presented in Fig. 4.7.

Figure 4.7 show a consistent increase in Pe_L with a decrease in Sc and it may be seen that the dependence is slight for the higher values of Sc (say for Sc of order 750 and above). At the lower end of the range of Pe_m investigated there seems to be a tendency for Pe_L to become independent of Sc, even if the values of D_L are still significantly above D_m. In the intermediate range, $100 < Pe_m < 5000$, values of Pe_L are very nearly constant, for each value of Sc. The convergence of the different series of points at about $Pe_m \cong 20$ seems to suggest that Pe_L is insensitive to Sc below this value of Pe_m, for the range of Sc presented.

A good additional test of the consistency of the data of Guedes de Carvalho and Delgado [61] is supplied by the plot in Fig. 4.8, where it may be seen that all the series of points converge at high Re, as would be expected for turbulent flow. The agreement with the data of Jacques and Vermeulen [79] and Miller and King [95], for cold water, is worth stressing.

Recently, some workers have measured axial dispersion for the flow of supercritical carbon dioxide through fixed beds and this provides important new data in the range $1.5 < Sc < 20$. However, the various authors fail to recognize the direct dependence of Pe_L on Sc. Catchpole et al. [27] represent their data and those of Tan and Liou [132] in a single plot (their Fig. 4.3) of Pe_L *vs*. Re. The majority of points are in the range of $1 < Re < 30$ and the data of both groups, together,

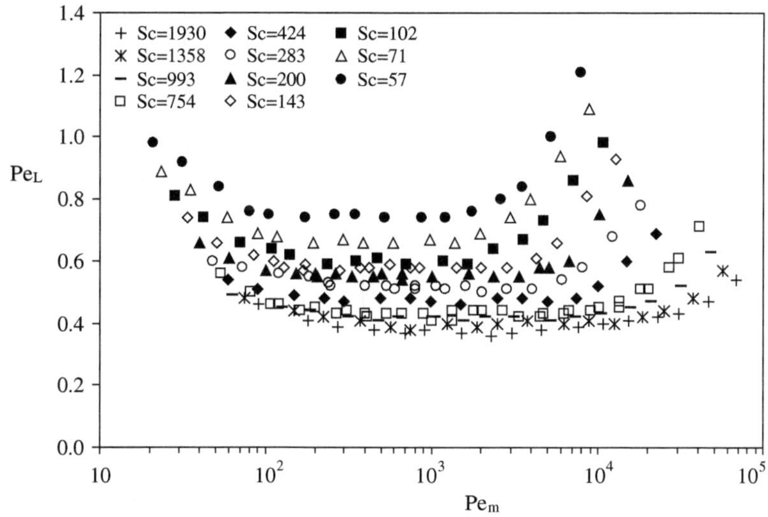

Fig. 4.7 Dependence of Pe_L on Pe_m for different values of Sc

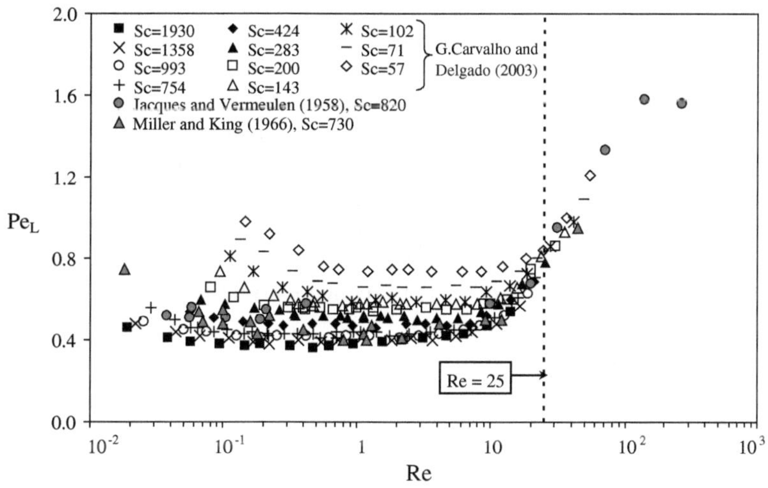

Fig. 4.8 Dependence of Pe_L on Re for different values of Sc

define a horizontal cloud with mid line at about $Pe_L \cong 0.8$, spreading over the approximate range $0.3 < Pe_L < 1.1$.

The data of Yu et al. [147] are for $0.01 < Re < 2$ and $2 < Sc < 9$. It is worth referring here that the modelling work of Coelho et al. [32] gives theoretical support to experimental findings for low Re, both for spherical and non-spherical particles. No influence of Sc on Pe_L is detected, but unfortunately the results are not very consistent, particularly in the range $1 < Pe_m < 20$, where the scatter is high and the values of Pe_L are much too low.

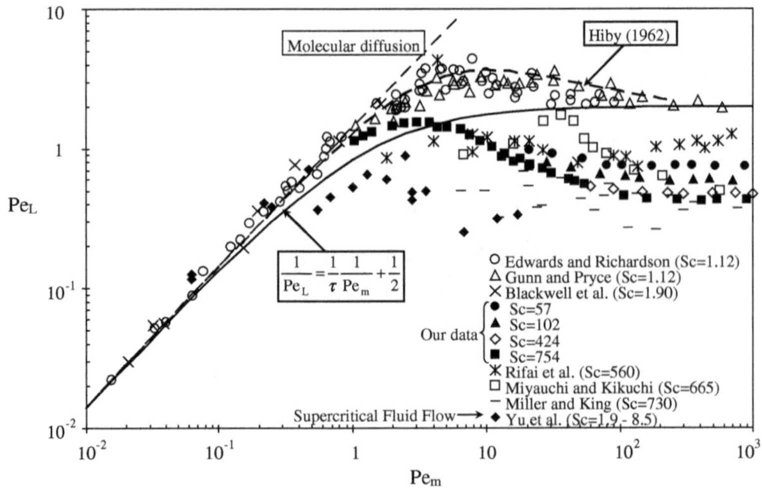

Fig. 4.9 Dependence of Pe_L on Pe_m for Stokes flow regime

Figure 4.9 shows that for low values of Pe_m (Stokes flow regime) there seems to be a tendency for Pe_L to become independent of Sc. The values of Pe_L reported by Miller and King [95], for $6 < Pe_m < 100$, are much too low; this may be because the particles used in most experiments are too small (particle sizes of 55 and 99 μm) and this is known to yield enhanced dispersion coefficients, possibly due to particle agglomeration (see [64, 75]). The data reported by Miyauchi and Kikuchi [97] and plotted in Fig. 4.9, for $6 < Pe_m < 300$, are higher than our experimental data.

There are considerable experimental difficulties in the measurement of axial dispersion in the liquid phase at small Reynolds number, because the usual method of obtaining low Reynolds number is to reduce particle size and this is known to yield enhanced dispersion coefficients.

4.4 Radial Dispersion

Generally, radial dispersion coefficients are measured in non-reactive conditions, because the rate of mass transfer, observed experimentally, is directly related to the coefficient of radial dispersion in the bed.

The most popular technique for the measurement of radial dispersion consists in feeding a continuous stream of tracer from a "point" source somewhere in the bed (usually along the axis, if there is one) and measuring the radial variation of tracer concentration at one or more downstream locations.

The first study of mass transfer by radial dispersion in gaseous systems was carried out by Towle and Sherwood [134]. The results presented were very important for packed bed dispersion because they showed that dispersion was not influenced by the tracer molecular weight.

Bernard and Wilhelm [14] reported the first measurements, in liquid systems, of experimental values of radial dispersion coefficients in packed beds of inerts by a Fickian model. The authors took into account the wall effect condition and their experiments suggested that for high values of Reynolds number the value of $\mathrm{Pe_T}$ is constant and between 11 and 13.

Baron [7] proposed a new model of radial dispersion in which a particle of tracer executes a simple random-walk displacement of $\pm 1/2$ particle diameter to give a transversal Peclet number between 5 and 13, when $\mathrm{Re} \rightarrow \infty$. The basis for this prediction is the random-walk theory, in which a statistical approach is employed. This method does not take into account effects of radial variations in velocity and void space. Latinen [87] extended the random-walk concept to three dimensions and predicted a value of 11.3, for $\mathrm{Pe_T}(\infty)$.

Klinkenberg et al. [81] solved Eq. 4.1 for anisotropic dispersion, but considered that dispersion occurs in an infinite medium. In the same work were considered the particulate cases of isotropic dispersion and axial dispersion neglected.

Plautz and Johnstone [105] used the equation derived by Wilson [145], for heat transfer, and suggested a $\mathrm{Pe_T}$ between 11 and 13, for $\mathrm{Re} \rightarrow \infty$. Fahien and Smith [51] assumed that for Reynolds numbers in the range between 40 and 100, the Peclet number is independent of fluid velocity and equal to 8. The authors were the first to consider that the tracer pipe can be of significant diameter compared to the diameter of the bed.

Dorweiler and Fahien [42] used the equation derived by Fahien and Smith [51] to study mass transfer in laminar and transient flows. The results showed that for $\mathrm{Re} < 200$, the Peclet number based on the radial dispersion coefficient is a linear function of the fluid velocity and for $\mathrm{Re} > 200$, at room temperature, the Peclet number is constant as also shown by Bernard and Wilhelm [14], Plautz and Johnstone [105] and Fahien and Smith [51]. The authors have demonstrated a difference in the Peclet number with radial position. The transversal Peclet number is constant from the axis to 0.8 times the radius and then rises near the wall.

Hiby and Schummer [76], and later Roemer et al. [111], presented the solution of the mass balance equation (Eq. 4.1), considering the tracer pipe to be of significant diameter compared to the diameter of the packed bed.

Saffman [114] considered the packed bed as a network of capillary tubes randomly orientated with respect to the main flow. At high Peclet number and at very long time, Saffman found that the dispersion never becomes truly mechanical, with zero velocity of the fluid at the capillary walls, the time required for a tracer particle to leave a capillary would become infinite as its distance from the walls goes to zero. The author proposed that $D_\mathrm{T} = (3/16)ud$ when $\mathrm{Re} \rightarrow \infty$, but this prevision of radial dispersion coefficient is higher than observed experimentally.

Hiby [75] and Blackwell [16] presented an experimental technique in which they divided the sampling region into two annular regions and calculated the transversal dispersion coefficient from the averaged concentrations of each of the two samples.

The experimental data points of Wilhelm [144] suggested that $\mathrm{Pe_T}(\infty) = 12$, for beds of closely sized particles, and this value is accepted for the majority of the investigators (ex: [15, 33, 66, 75, 144]).

Roemer et al. [111] studied radial mass transfer in packed beds at low flow rates, Re <100. The authors considered the tracer pipe to be of significant diameter compared to the diameter of the bed ("finite source" model) and axial and radial dispersion are equal. In this work the authors compared the values of Pe_T obtained with two methods ("instantaneous finite source" and "point source") and concluded that the values of Pe_T obtained with the "point source" method were 10% less that the values obtained with the "instantaneous finite source" method. The authors estimated that the neglecting the axial dispersion in calculations of D_T, for low values of Reynolds numbers, can cause errors of 10%.

Coelho and Guedes de Carvalho [33] developed a new experimental technique, based on the measurement of the rate of dissolution of planar or cylindrical surfaces, buried in the bed of inert particles and aligned with the flow direction. This alternative technique is simple to use, allows the determination of the coefficient of radial dispersion in packed beds over a wide range of flow rates, and it is easily adaptable to work over a range of temperatures above ambient, as shown by Guedes de Carvalho and Delgado [63] and Delgado and Guedes de Carvalho [40].

In recent years, nuclear magnetic resonance has been used to determine both diffusion and dispersion coefficients (e.g. [8, 57]), with significant advantages, but this technique were limited to low fluid velocities.

It is important to remember that, at high Reynolds numbers, the main mechanism of radial dispersion is the fluid deflection caused by deviations in the flow path caused by the particles in the bed (axial dispersion is caused by differences in fluid velocity in the flow), i.e., dispersion is caused by hydrodynamic mechanisms (macroscopic scale) and not by molecular diffusion (Brownian motion).

The result is a poor mixture at the "microscopic scale". In fact, there are detected different values of solute concentration over a distance of the order of a particle diameter or less, what explains the convenience of use of an efficient averaging procedure [66]. This is probably one of the reasons that explain the difference observed in some experimental results of dispersion (see Fig. 4.10). Gunn and Pryce [68] showed that the standard deviation without repacking in the measurement of Pe_T was 5%, while when the bed was repacked each time of measurement, the standard deviation found was 15%.

4.4.1 Parameters Influencing Radial Dispersion: Porous Medium

4.4.1.1 Length of the Packed Column

Han et al. [69] showed that values of the radial dispersion coefficient, for uniform size packed beds, measured at different positions in the bed are not a function of bed location, *i.e.*, they observed no time dependent behaviour for radial dispersion, because radial dispersion is caused by mechanical mechanism alone.

An important aspect to be considered, as a check on the experimental method of Coelho and Guedes de Carvalho [33], is the influence of the length of the test

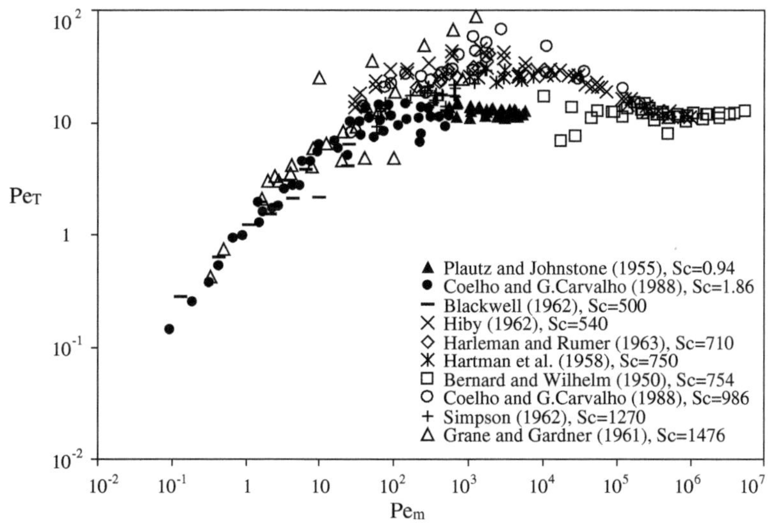

Fig. 4.10 Some experimental data points for radial dispersion in liquid systems and gaseous systems

cylinder on the measured value of D_T. In reality, the two variables are independent, provided that the criterion given by Eq. 4.46 is satisfied (see Fig. 4.11)

$$\frac{L}{d} \geq 0.62 \left(\frac{ud}{D_m}\right) \tag{4.46}$$

4.4.1.2 Ratio of Column Diameter to Particle Diameter

Several investigators, like Fahien and Smith [51], Latinen [87] and Singer and Wilhelm [125], have studied the wall effect on radial dispersion coefficient. The experiments suggested that in a packing structure characterized by significant variations of void fraction in radial direction, up to distance of about two particle diameters from the wall, a non-uniform radial velocity profile is induced, with a maximum just near the wall. As result, wall effects occur due large voidage fluctuations near the wall. The above investigators also showed that the increase in radial dispersion in the laminar region would be the same order of magnitude as in the turbulent region.

4.4.1.3 Particle Size Distribution

Eidsath et al. [48] studied the effect of particle size distribution on dispersion. As the ratio of particle diameter went from a value of 2 to 5, the radial dispersion decreased by a factor of 3, but perhaps the results were a cause of the simple

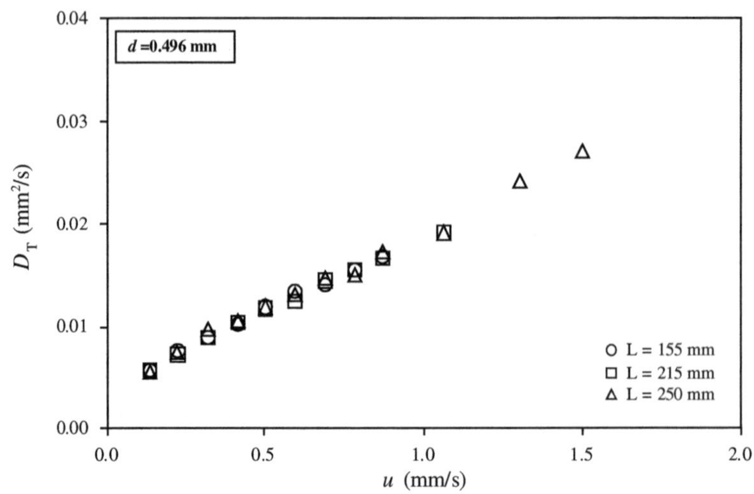

Fig. 4.11 Effect of length of soluble cylinder on the measurement of radial dispersion

geometry employed in these computations (packed bed of cylinders). Steady-state measurements of radial dispersion reported by Han et al. [69], with the same void fraction and mean particle diameter, but different particle size range (ratio of maximum to minimum particle diameter equal to 2.2 and 7.3), showed that there was no evidence to indicate a change in radial dispersion with particle size distribution (see Fig. 4.12a).

The effect of a distribution of particle sizes within the bed, on the radial dispersion coefficient, may be assessed from Guedes de Carvalho and Delgado [63]. In particular, lot D was prepared by carefully blending lots B and E in a proportion of 1:1 (by weight). In Fig. 4.12b, dispersion data obtained with the mixed lot are seen to fall in between the data for the original separate lots, as might be expected. Figure 4.12a shows that in a plot of D_T/D_m *vs.* Pe_m, the data for the three lots fall along the same line, when *d* (in Pe_m) is taken to represent the average particle size in the bed.

4.4.1.4 Particle Shape

The effect of particle shape on the radial dispersion coefficient has been given attention by several investigators both for gaseous and liquid systems. England and Gunn [49] measured the dispersion of argon in beds of solid cylinders and beds of hollow cylinders and have concluded that D_T tend to be greater with packs of hollow cylinders than with packs of solid cylinders, and these results were greater than obtained with packs of spherical particles (see Fig. 4.13).

The same conclusion, in liquid systems, was been obtained by Hiby [75], who used packed beds of glass spheres and Rachig rings, and Bernard and Wilhem [14],

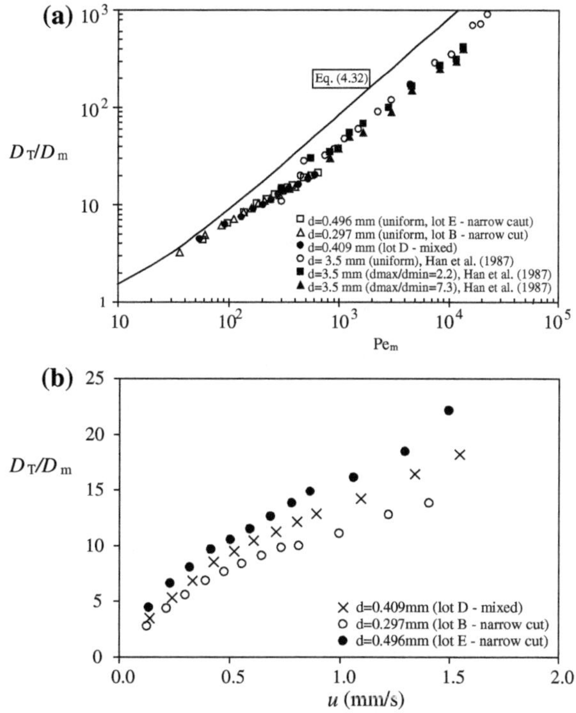

Fig. 4.12 Effect of particle size distribution on radial dispersion. **a** D_T/D_m *vs.* Pe_m; **b** D_T/D_m *vs.* u.

who used packed beds of cubes, cylinders and glass spheres. Figure 4.13 shows that the radial dispersion coefficient tends to be greater in packed beds of non-spherical particles.

However, Blackwell [16], List [91], Guedes de Carvalho and Delgado [63] and others reported experiments with packed beds of sand and showed that D_T obtained with glass ballotini are very close to those for sand (not pebble or gravel) and the conclusion seems to be that particle shape has only a small influence on lateral dispersion, for random packings of "isometric" particles.

4.4.2 Parameters Influencing Radial Dispersion: Fluid Properties

4.4.2.1 Viscosity and Density of the Fluid

The effect of fluid densities and viscous forces on radial dispersion has been studied by Grane and Garner [58] and Pozzi and Blackwell [106]. They concluded that when a fluid is displaced from a packed bed by a less viscous fluid, the viscous forces create an unstable pressure distribution and the less viscous fluid will penetrate the medium in the form of fingers, unless the density has an opposing effect.

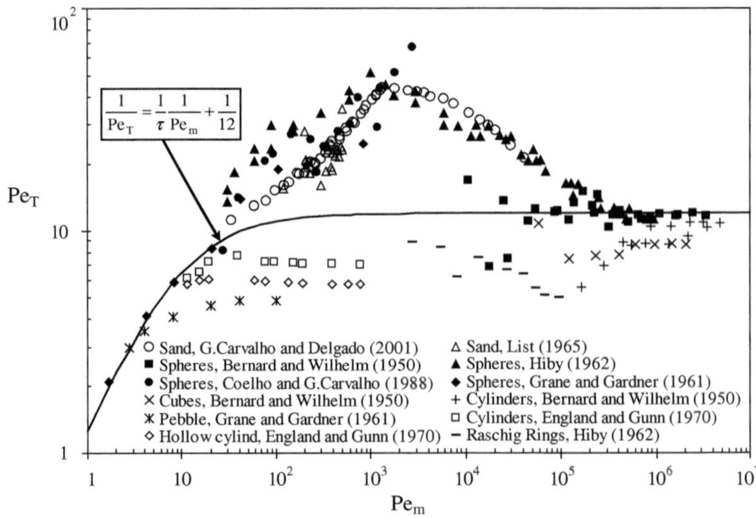

Fig. 4.13 Effect of particle shape on radial dispersion

4.4.2.2 Fluid Velocity

For very low fluid velocities, u, dispersion is the direct result of molecular diffusion, with $D_T = D'_m$. As the velocity of the fluid is increased, the contribution of convective dispersion becomes dominant over that of molecular diffusion and D_T becomes less sensitive to temperature. According to several authors (see [15, 33, 66, 75, 144]) $D_T \to ud/\mathrm{Pe_T}(\infty)$, for high enough values of u, where d is particle size and $\mathrm{Pe_T}(\infty) \cong 12$ for beds of closely sized particles. Assuming that the diffusive and convective components of dispersion are additive, the same authors suggest that $D_T = D'_m + ud/K$, which may be written in dimensionless form as

$$\frac{D_T}{D_m} = \frac{1}{\tau} + \frac{1}{12}\frac{ud}{D_m} \quad \text{or} \quad \frac{1}{\mathrm{Pe_T}} = \frac{1}{\tau}\frac{\varepsilon}{\mathrm{ReSc}} + \frac{1}{12} \tag{4.47}$$

This equation has been shown (see [33]) to give a fairly accurate description of radial dispersion in gas flow through packed beds, but it is not appropriate for the description of dispersion in liquids, over an intermediate range of values of ud/D_m, as pointed out by several of the authors mentioned above.

Figure 4.14a–b shows that the value of the radial dispersion coefficient is seen to increase with fluid velocity and comparison between the two plots shows that D_T also increases with particle size.

Data on dispersion in randomly packed beds of closely sized, near spherical particles, lend themselves to simple correlation by means of dimensional analysis. Making use of Buckingham's theorem it may therefore be concluded that

$$\frac{D_{\rm T}}{D_{\rm m}} = \Phi\left(\frac{ud}{D_m}, \frac{\mu}{\rho D_m}\right) \quad \text{or} \quad {\rm Pe_L} = \Phi({\rm Re, Sc}) \tag{4.48}$$

and it is useful to make $\mathrm{Pe_m} = ud/D_\mathrm{m}$ and $\mathrm{Sc} = \mu/\rho D_\mathrm{m}$.

4.4.2.3 Fluid Temperature (or Schmidt Number)

The dependence of D_T on liquid properties and velocity is best given in plots of $\mathrm{Pe_T}$ *vs.* $\mathrm{Pe_m}$, for different values of Sc. Not surprisingly, Fig. 4.15 shows that the variation of $\mathrm{Pe_T}$ with $\mathrm{Pe_m}$ gets closer to that for gas flow as the value of Sc is decreased. For the lowest Sc tested ($\mathrm{Sc} = 54$; $T = 373$ K), $\mathrm{Pe_T}$ does not differ by more than 30% from the value given by Eq. 4.47, with $\mathrm{Pe_T}(\infty) = 12$, over the entire range of $\mathrm{Pe_m}$. However, for the higher values of Sc, the experimental values of $\mathrm{Pe_T}$ may be up to four times the values given by Eq. 4.47.

Delgado and Guedes de Carvalho [40] had studied the dependence of $D_\mathrm{T}/D_\mathrm{m}$ on Sc, up to $\mathrm{Pe_m} \cong 1350$, and they reported a smooth increase in $D_\mathrm{T}/D_\mathrm{m}$ with $\mathrm{Pe_m}$, for all values of Sc. But the data in Fig. 4.15 show that there is a sudden change in the trend of variation of $\mathrm{Pe_T}$ with $\mathrm{Pe_m}$, somewhere above $\mathrm{Pe_m} \cong 1,350$, a maximum being reached in the approximate range $1,400 < \mathrm{Pe_m} < 1,800$ (depending on Sc). The fact that the change in trend corresponds to a much enhanced increase in D_T (*i.e.* a decrease in $\mathrm{Pe_T}$), in response to a small increase in u (*i.e.* in $\mathrm{Pe_m}$), strongly suggests a connection with the transition from laminar to turbulent flow in the interstices of the packing. The plot of $\mathrm{Pe_T}$ *vs.* Re, shown in Fig. 4.16, seems to support this view, since the maxima in $\mathrm{Pe_T}$ are reached for $0.3 < \mathrm{Re} < 10$ (depending on Sc) and this is the approximate range of values of Re for the transition from laminar to turbulent flow. The range $1 < \mathrm{Re} < 10$ is often indicated for that transition (see for example [9]), but Scheidegger [116] as giving $\mathrm{Re} = 0.1$ for the lower limit of that transition.

The plot in Fig. 4.16 also suggests that "purely mechanical" fluid dispersion will be observed above about $\mathrm{Re} = 100$; this value is estimated as the convergence of the data points for liquids with the line representing Eq. 4.47. Figure 4.17 shows the data reported by most other authors (all for $\mathrm{Sc} \geq 540$) in a plot of $\mathrm{Pe_T}$ *vs.* $\mathrm{Pe_m}$. With the exception of the data of Hoopes and Harleman [78] and some of the points of Grane and Gardner [58] and Bernard and Wilhelm [14], general agreement is observed with Guedes de Carvalho and Delgado [62] data for high Sc.

4.5 Dispersion in Packed Beds Flowing by Non-Newtonian Fluids

Hilal et al. [77], Edwards and Helail [46], Payne and Paker [101] and Wen and Yin [142] reported results of axial dispersion coefficients for the flow of two polymer solutions through a packed bed and their results were similar to the corresponding

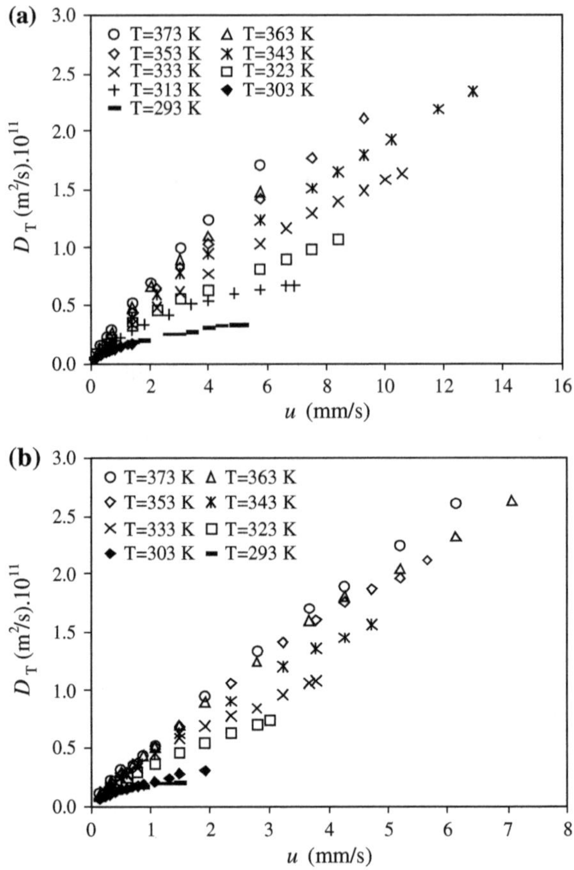

Fig. 4.14 Variation of radial dispersion coefficient with fluid velocity. **a** sand size $d = 0.297$ mm; **b** sand size $d = 0.496$ mm

Newtonian results (see Table 4.1). Wen and Fan [140] correlate the previous results for packed beds with the following expression:

$$\mathrm{Pe}\left(=\frac{Ud}{D_{\mathrm{L}}}\right) = 0.2 + 0.011\mathrm{Re}_n^{0.48} \,\text{with}\, \mathrm{Re}_n = \frac{\rho d^n U^{2-n}}{m} \tag{4.49}$$

where m is the power law consistency coefficient. Note that Eq. 4.49 for $n = 1$ (Newtonian fluids) reduces to the correlation obtained by Chung and Wen [31], for Newtonian fluid through packed beds.

The only study on the influence of Non-Newtonian fluid in radial dispersion coefficients is reported by Hassell and Bondi [73]. who showed that the quality of mixing deteriorate with increasing viscosity.

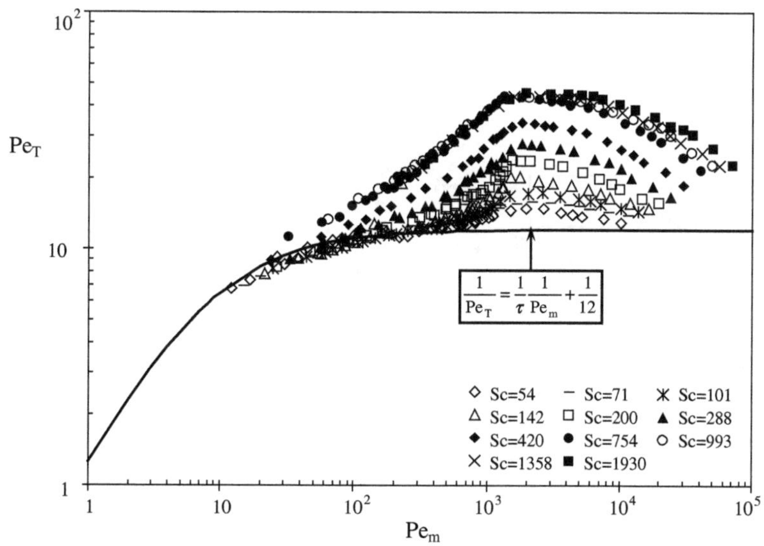

Fig. 4.15 Dependence of Pe_T on Pe_m for different values of Sc

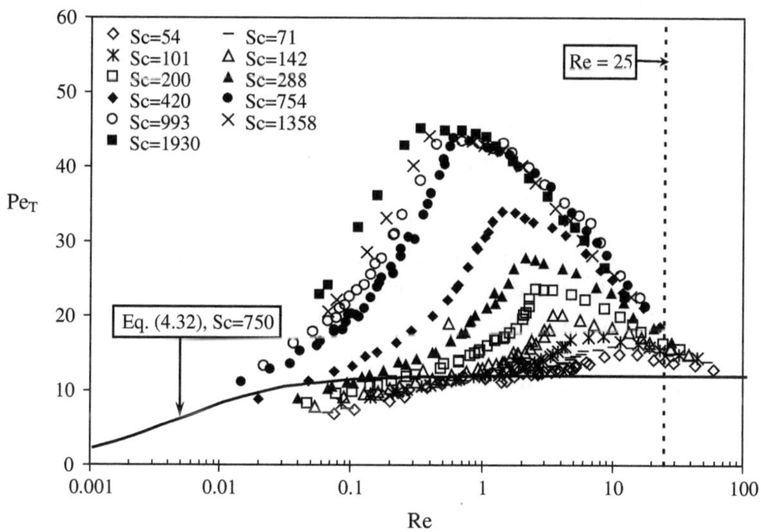

Fig. 4.16 Dependence of Pe_T on Re for different values of Sc

4.6 Correlations

Many equations have been proposed to correlate dispersion experimental data in porous media, and the most used are the equations proposed by Bear and Verruijt [10] in groundwater works and the equations proposed by Koch and Brady [83].

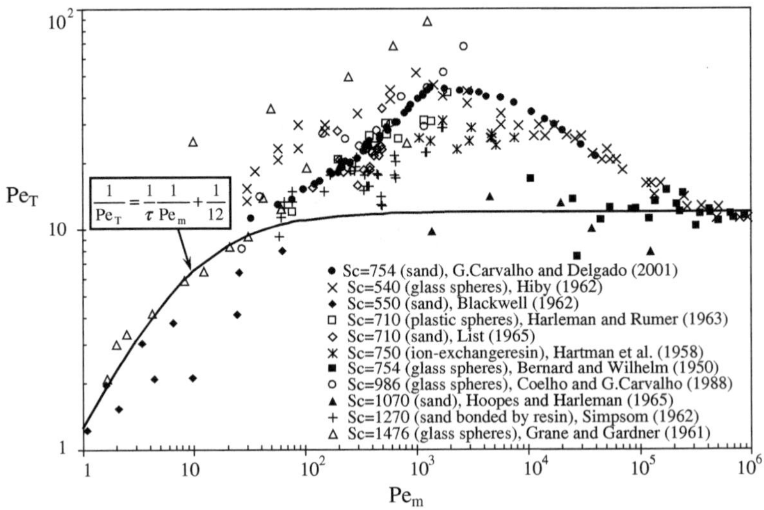

Fig. 4.17 Comparison between our data points and the results of other authors for $\mathrm{Sc} \geq 540$

The coefficients of radial and longitudinal dispersion are not linear function of velocity and expressions with the form $D_\mathrm{L} = D'_\mathrm{m} + \alpha_\mathrm{L} u^n$ or $D_\mathrm{T} = D'_\mathrm{m} + \alpha_\mathrm{T} u^n$ are suggested. The coefficients α_T and α_L are the radial and longitudinal dispersivities, respectively, of the porous medium in the direction of transport and n is an empirically constant, $n = 1 - 2$ [53]. In most applications, the exponent, n, is assumed to be unity, i.e., dispersion coefficient is assumed to be a linear function of fluid velocity. However, n may be greater than unity in many situations [10]. Also, the dispersivity, α_L, is probably scale dependent with larger values for α_L being associated with greater transport distances [104]. For example, values of α_L reported from the results of field studies may be as much as four or six orders of magnitude greater than the corresponding laboratory measured values which commonly are found to range between 0.1 and 10 mm [53]. Ratios of $\alpha_\mathrm{L}/\alpha_\mathrm{T}$ of 5:1–100:1 have been reported in the literature [10].

Some of most referred works were developed by Fried and Combarnous [54] and Bear and Verruijt [10, p. 166]; the authors showed the existence of five dispersion regimes, in unconsolidated porous media. Sahimi [115] and Marsily [92] analyze the data compiled by Fried and Combarnous [54] to characterize longitudinal dispersion in five dispersion regimes and radial dispersion in four dispersion regimes and a hold-up dispersion. The (1) pure molecular diffusion regime, for very low fluid velocity, is represented by Koplik et al. [84]

$$\frac{D_\mathrm{L}}{D_\mathrm{m}} = \frac{D_\mathrm{T}}{D_\mathrm{m}} = \frac{1}{\tau} \tag{4.50}$$

The (2) superposition regime $(0.3 < \mathrm{Pe_m} < 5)$ not quantify by an equation. In this zone the effect of molecular diffusion and dispersion are of the same order of magnitude. The (3) predominant mechanical dispersion $(5 < \mathrm{Pe_m} < 300)$:

Table 4.1 Summary of the previous work with experimental data on axial and radial dispersion of non-newtonian fluids in packed beds

Reference	Experimental method	Packed bed	ε	d (mm)	L (mm)	D (mm)	Re_γ	Sc
Wen and Yin [142]	Pulse response (H_2O+Polyox 301)—$Na_2C_{20}H_{10}O_5$ ($n = 0.90$ and 0.81)	Glass spheres	0.40 0.50	4.76 14.3	1066.8	50.8	7–800	$\cong 10^6$
Payne and Parker [101]	Step function (H_2O+Polyox 301)—blue dye ($n = 0.95$ and 0.52)	Glass spheres	0.365	0.374	595	38.1	9×10^{-6}–1.5×10^{-3}	8×10^6–10^9
Edwards and Helail [46]		Glass spheres						
Hilal et al. [77]	Frequency response (Potassium ferricyanide and ferrocyanid +CMC sodium salt+Na_2CO_3)	Glass spheres parallelepiped particles	0.39	2–51.05 × 5.0 ($d = 0.42$)				
Hassel and Bondi [73]		Glass spheres		5.0				

$$\frac{D_L}{D_m} = \frac{1}{\tau} + 0.5\mathrm{Pe}_m^{1.2} \tag{4.51a}$$

$$\frac{D_T}{D_m} = \frac{1}{\tau} + 0.025\mathrm{Pe}_m^{1.1} \tag{4.51b}$$

The (4) pure mechanical dispersion $\left(300 < \mathrm{Pe_m} < 10^5\right)$ represented by

$$\frac{D_L}{D_m} = \frac{1}{\tau} + (1.8 \pm 0.4)\mathrm{Pe}_m \tag{4.52a}$$

$$\frac{D_T}{D_m} = \frac{1}{\tau} + 0.025\mathrm{Pe}_m \tag{4.52b}$$

In this zone (yet in the range of validity of Darcy's law) the effect of molecular diffusion is negligible. The (5) dispersion out of Darcy domain $\left(\mathrm{Pe_m} > 10^5\right)$ not quantify by an equation. This is the region in which the effects of inertia and turbulence cannot be neglected.

The (6) holdup dispersion [83]:

$$\frac{D_L}{D_m} = \frac{D_T}{D_m} \approx \mathrm{Pe}_m^2 \tag{4.53}$$

Probably one of the most important studies in dispersion topic was presented by Saffman [113, 114]. These works are the most detailed analysis of dispersion in porous media, but the correlation proposed for axial and radial dispersion diverge slightly from the existing experimental data, in the literature (see Fig. 4.18). The author proposed the following equation, for longitudinal dispersion:

$$\frac{D_L}{D'_m} = \frac{\mathrm{Pe}'_m}{6}\left[\ln\left(\frac{3}{2}\tau\mathrm{Pe}'_m\right) - \frac{1}{4}\right] \quad \text{valid for} \quad \mathrm{Pe}'_m \gg 1 \tag{4.54}$$

Another important study was presented by Koch and Brady [83] who showed that the ratio of the dispersion coefficient to the molecular diffusivity is only a function of the product ReSc. However, experimental measurements show that the dispersion Peclet group is a function of both Reynolds and Schmidt groups as $\mathrm{Pe} = f(\mathrm{Re},\ \mathrm{Sc})$ and not $\mathrm{Pe} = f(\mathrm{ReSc})$. For Stokes flow through a random packed bed of spheres, Koch and Brady [83] derived analytical expressions for the longitudinal and radial dispersion coefficients. Since the solid phase is not permeable to the tracer, for $\mathrm{Pe}'_m > 1$, D_L/D'_m and D_T/D'_m are given by:

$$\frac{D_L}{D'_m} = 1 + \frac{3}{4}\frac{\mathrm{Pe}'_m}{2} + \frac{\pi^2}{6}(1-\varepsilon)\frac{\mathrm{Pe}'_m}{2}\ln\left(\frac{\mathrm{Pe}'_m}{2}\right) \tag{4.55}$$

$$\frac{D_T}{D'_m} = 1 + \frac{63\sqrt{2}}{320}\sqrt{1-\varepsilon}\frac{\mathrm{Pe}'_m}{2} \tag{4.56}$$

The works of Saffman [113, 114] and Koch and Brady [83] have shown that $D_L/D'_m \sim Pe'_m \ln(Pe'_m)$ and $D_T/D'_m \sim Pe'_m$. The logarithmic dependence results from the fact that velocity of fluid at the walls is zero. Different Peclet number contributions to dispersion were identified: hold-up in a closed circulation region $\left(Pe'_m 2\right)$, particle boundary layer $\left(Pe'_m \ln(Pe'_m)\right)$ and random velocity field contribution $\left(Pe'_m\right)$. The agreement with the experimental results of Fried and Combarnous [54] was very good for both longitudinal and radial dispersivities. Saffman [113] modelled random porous media with randomly oriented capillary tubes and found Fickian dispersion at long times with dispersivity approaching a $Pe'_m \ln(Pe'_m)$ behaviour.

It's important to enhance the fact that in literature, it was possible to find a large number of empirical correlations. For example, Fetter [52] used the experimental data obtained by Carberry and Bretton [25], Raimondi et al. [108], Blackwell et al. [17] and Rosenberg [112] to conclude that longitudinal dispersion coefficient is well represented by the fitted curve:

$$\frac{D_L}{D'_m} = 1 + 1.75 Pe'_m \quad \text{valid for } Pe'_m < 50\tau \tag{4.57}$$

and the data of Blackwell [16] and Grane and Gardner [58], to suggest the following equation for radial dispersion:

$$\frac{D_T}{D'_m} = 1 + 0.055 Pe'_m \tag{4.58}$$

Gunn [65] admitted the existence of two regions in the packing, one of fast flowing and the other of nearly stagnant fluid, to deduce the following expression for the axial dispersion coefficient in terms of probability theory

$$\begin{aligned} \frac{1}{Pe_L} = {} & \frac{\varepsilon Pe_m}{4\alpha_1^2(1-\varepsilon)}(1-p)^2 + \left[\frac{\varepsilon Pe_m}{4\alpha_1^2(1-\varepsilon)}\right]^2 p(1-p)^3 \\ & \times \left\{\exp\left[-\frac{4(1-\varepsilon)\alpha_1^2}{p(1-p)\varepsilon\, Pe_m}\right] - 1\right\} + \frac{1}{\tau\, Pe_m} \end{aligned} \tag{4.59}$$

where α_1 is the first zero of equation $J_0(U) = 0$ and p is defined, for a packing of spherical particles, by

$$p = 0.17 + 0.20 \times \exp\left(-\frac{24}{Re}\right) \quad \text{for spheres, } \tau = \sqrt{2} \tag{4.60a}$$

$$p = 0.17 + 0.20 \times \exp\left(-\frac{24}{Re}\right) \quad \text{for solid cylinders, } \tau = 1.93 \tag{4.60b}$$

$$p = 0.17 + 0.20 \times \exp\left(-\frac{24}{Re}\right) \quad \text{for hollow cylinders, } \tau = 1.8 \tag{4.60c}$$

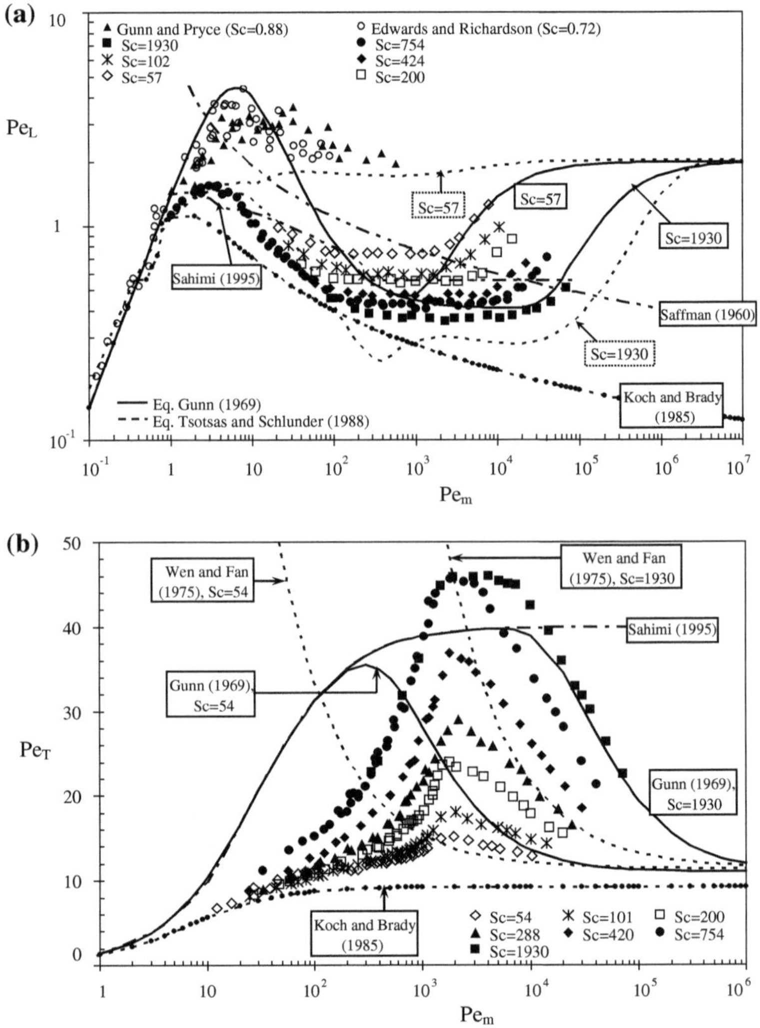

Fig. 4.18 Comparison between experimental data and correlations presented in the literature

Tsotsas and Schlunder [135] deduced an alternative correlation for the prediction of $\mathrm{Pe_L}$. The authors defining two zones in a simple flow model consisting of a fast stream (central zone in the model capillary) and a stagnant fluid, but the mathematical expressions associated with it are a little cumbersome,

$$\frac{1}{\mathrm{Pe_L}} = \frac{1}{\tau}\left[\frac{1}{\mathrm{Pe}_{z,1}} + \frac{1}{\mathrm{Pe}'_{\mathrm{m}}}\left(1 - \xi_{\mathrm{c}}^{2}\right)\right] + \frac{1}{32}\left(\frac{D_{\mathrm{c}}}{d}\right)^{2}\left[\mathrm{Pe}_{\mathrm{r},1}\xi_{\mathrm{c}}^{2} f_1(\xi_{\mathrm{c}}) + \mathrm{Pe}'_{\mathrm{m}} f_2(\xi_{\mathrm{c}})\right] \tag{4.61}$$

where the axial and radial Peclet number of the fast stream is

$$\frac{1}{\mathrm{Pe}_{z,1}} = \frac{1}{\mathrm{Pe}'_1} + \frac{1}{1.14(1 + 10/\mathrm{Pe}'_1)} \tag{4.62a}$$

$$\frac{1}{\mathrm{Pe}_{r,1}} = \frac{1}{\mathrm{Pe}'_1} + \frac{1}{8} \tag{4.62b}$$

$$\mathrm{Pe}'_1 = \frac{u_1 d}{D'_\mathrm{m}} \tag{4.62c}$$

and $u_1 = u/\xi_c^2$ is the interstitial velocity of the fast stream, with ξ_c (the dimensionless position of the velocity jump, *i.e.* the ratio between the radius of the zone of high velocities and the radius of packed bed) equal to

$$\mathrm{Re} \le 0.1 \rightarrow \xi_c = 0.2 + 0.21\exp(2.81y) \tag{4.62d}$$

$$\mathrm{Re} \ge 0.1 \rightarrow \xi_c = 1 - 0.59\exp[-\mathrm{f}(y)] \tag{4.62e}$$

with

$$y = \log(\mathrm{Re}) + 1 \tag{4.62f}$$

$$\mathrm{f}(y) = y\left(1 - 0.274y + 0.086y^2\right) \tag{4.62g}$$

Finally, the distributions functions $f_1(\xi_c)$ and $f_2(\xi_c)$ are defined by:

$$f_1(\xi_c) = \left(1 - \xi_c^2\right)^2 \tag{4.62h}$$

$$f_2(\xi_c) = 4\xi_c^2 - 3 - 4\ln(\xi_c) - \xi_c^4 \tag{4.62i}$$

In Fig. 4.18a, the lines corresponding to the correlations of Gunn [65] and of Tsotsas and Schlunder [135] are represented, for the higher and lower values of Sc in our experiments (Sc = 57 and 1,930), as well as for gas flow (Sc = 1). It may be seen that the correlation of Gunn [65] is not sensitive to changes in Sc, for $\mathrm{Pe_m} < 10^3$, and the correlation of Tsotsas and Schlunder [135] is much too sensitive to variations in Sc; however, this correlation describes dispersion in gas flow with good accuracy.

In this context it is interesting to consider, for radial dispersion, the predicting accuracy of some alternative empirical correlations that have been proposed to represent the experimental data in liquid flow, as the equation of Gunn [65]:

$$\frac{1}{\mathrm{Pe_T}} = \frac{1}{\mathrm{Pe_f}} + \frac{1}{\tau}\frac{\varepsilon}{\mathrm{ReSc}} \tag{4.63}$$

where the fluid-mechanical Peclet number, $\mathrm{Pe_f}$, is defined by,

$$\mathrm{Pe_f} = 40 - 29\mathrm{e}^{-7/\mathrm{Re}} \quad \text{for spheres, } \tau = \sqrt{2} \tag{4.64a}$$

$$\mathrm{Pe_f} = 11 - 4\mathrm{e}^{-7/\mathrm{Re}} \quad \text{for solid cylinders, } \tau = 1.93 \tag{4.64b}$$

$$\mathrm{Pe_f} = 9 - 3.3\mathrm{e}^{-7/\mathrm{Re}} \quad \text{for hollow cylinders, } \tau = 1.8 \tag{4.64c}$$

And the empirical equation proposed by Wen and Fan [141],

$$\mathrm{Pe_T} = \frac{17.5}{\mathrm{Re}^{0.75}} + 11.4 \quad (\text{for high values of } \mathrm{Pe_m}) \tag{4.65}$$

In Fig. 4.18b, the lines corresponding Gunn [65] and Wen and Fan [141] correlations are represented for the two extreme values of Schmidt observed in our experiments (Sc = 54 and 1,930); comparison with the experimental points shows that the correlations are very inadequate over significant ranges of $\mathrm{Pe_m}$.

In conclusion, we can say that it was possible to find a large number of equations in the literature that have been proposed to correlate dispersion experimental data in porous media. However, the vast amount of data available is obtained mostly for air and water at room temperature (the influence of Schmidt number is not taking into account).

4.6.1 New Correlations: Axial Dispersion

In the limit of very low fluid velocity, dispersion is determined solely by molecular diffusion, with $D_\mathrm{L} = D_\mathrm{m}/\tau$ (τ being the tortuosity factor for diffusion). At high fluid velocities, dispersion is purely "fluid mechanical" (see [144]), with $D_\mathrm{L} = ud/\mathrm{Pe_L}(\infty)$, where u is the interstitial fluid velocity and $\mathrm{Pe_L}(\infty) \cong 2$ for gas or liquid flow through beds of (approximately) isometric particles, with diameter d. A common approximation for the intermediate range of fluid velocities is to assume that the effects of molecular diffusion and fluid mechanical dispersion are additive and the resulting expression is given by

$$D_\mathrm{L} = D'_\mathrm{m} + ud/\mathrm{Pe_L}(\infty) \tag{4.66}$$

where $D'_\mathrm{m} = D_\mathrm{m}/\tau$. This equation is expected to give the correct asymptotic behaviour at high and low values of $\mathrm{Pe_m}$. In the case of gas flow, see Fig. 4.19, significant deviations are observed in the range $0.6 < \mathrm{Pe_m} < 60$; the experimental values of $\mathrm{Pe_L}$ are generally higher than predicted by Eq. 4.66, with $\mathrm{Pe_L}(\infty) \cong 2$. Several equations have been proposed to represent the data in this intermediate range and the equation presented by Hiby [75],

$$\frac{D_\mathrm{L}}{D'_\mathrm{m}} = 1 + \frac{0.65\mathrm{Pe}'_\mathrm{m}}{1 + 7\sqrt{\tau/\mathrm{Pe}'_\mathrm{m}}} \quad (\text{valid for } \mathrm{Re} < 100) \tag{4.67}$$

is shown to fit the data points reasonably well.

For most gaseous mixtures, far from the critical point, the value of Sc is close to unity, whereas for most solutes in cold water, it is $550 < \mathrm{Sc} < 2,000$. In recent years, data on longitudinal dispersion have been made available for values of Sc between these ranges, such data were obtained with either supercritical carbon dioxide ($1.5 < \mathrm{Sc} < 20$) or heated water ($55 < \mathrm{Sc} < 550$).

Some workers have measured axial dispersion for the flow of supercritical carbon dioxide through fixed beds and this provides important new data in the range $1.5 < \mathrm{Sc} < 20$ (see Fig. 4.19). The experiments of Catchpole et al. [27] were performed in the range $8 < \mathrm{Sc} < 20$, but the individual values of Sc, for the data points represented, are not given by the authors. One would expect that the data of Catchpole et al. [27] would lie somewhere between the series of points for Sc = 57 and the line corresponding to Eq. 4.67. It turns out that values of $\mathrm{Pe_L}$ are a little lower than expected, possibly as a result of low accuracy; the scatter is certainly very pronounced. The data of Tan and Liou [132] represents some 90 data points by those authors, for $0.3 < \mathrm{Re} < 135$ and $1.5 < \mathrm{Sc} < 3$; the points are more or less evenly distributed, meaning that scatter is very significant.

The data of Yu et al. [147] are for $0.01 < \mathrm{Re} < 2$ and $2.0 < \mathrm{Sc} < 9.1$. Unfortunately they are not very consistent, particularly in the range $1 < \mathrm{Pe_m} < 20$, where the scatter is high and the values of $\mathrm{Pe_L}$ are much too low. The experiments of Ghoreishi and Akgermanb [56], again for the flow of supercritical carbon dioxide, are for $0.1 < \mathrm{Re} < 0.3$ and $3.3 < \mathrm{Sc} < 5.8$. No influence of Sc on $\mathrm{Pe_L}$ is detected, but this is what might be expected, considering that the values of Re are generally very low, with the consequence that the points are partly in the range where dispersion is totally determined by molecular diffusion.

For the case of liquid flow in a porous media, our group used the division in five dispersion regimes to obtain the expressions presented below.

(1) Diffusion regime (valid for $\mathrm{Pe_m} < 0.1$):

$$\frac{D_\mathrm{L}}{D'_\mathrm{m}} = 1 \tag{4.68}$$

(2) Predominant diffusional regime (valid for $0.1 < \mathrm{Pe_m} < 4$) :

$$\frac{D_\mathrm{L}}{D'_\mathrm{m}} = \frac{\mathrm{Pe}'_\mathrm{m}}{0.8/\mathrm{Pe}'_\mathrm{m} + 0.4} \tag{4.69}$$

with a an average relative deviation lesser than 14%.

(3) Predominant mechanical dispersion (valid for $4 < \mathrm{Pe_m}$ and $\mathrm{Re} < 10$) :

$$\frac{D_\mathrm{L}}{D'_\mathrm{m}} = \frac{\mathrm{Pe}'_\mathrm{m}}{\sqrt{18\,\mathrm{Pe}_\mathrm{m}^{-1.2} + 2.35\,\mathrm{Sc}^{-0.38}}} \tag{4.70}$$

with a deviation lesser than 11%, over the entire range of Pe'_m and Sc.

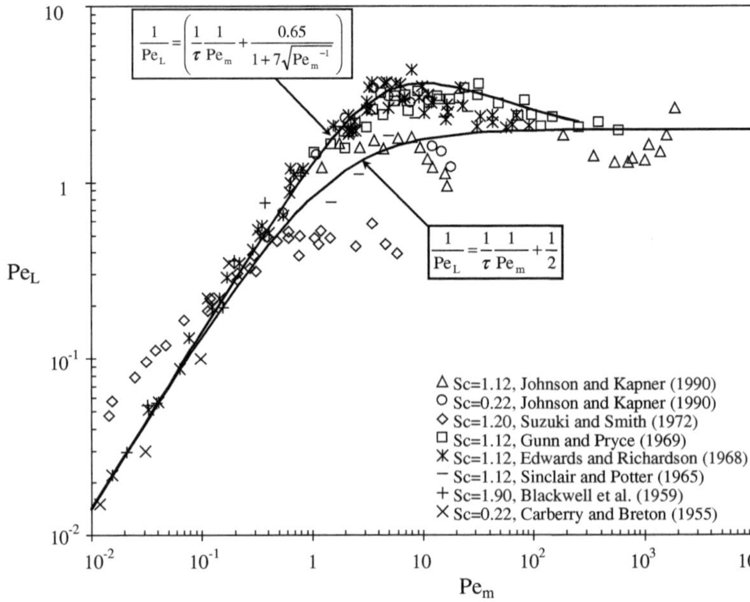

Fig. 4.19 Axial dispersion in gas flow

(4) Pure mechanical dispersion (valid for $10 < \text{Re}$ and $\text{Pe}_\text{m} < 10^6$) :

$$\frac{D_\text{L}}{D'_\text{m}} = \frac{\text{Pe}'_\text{m}}{25\text{Sc}^{1.14}/\text{Pe}'_\text{m} + 0.5} \tag{4.71}$$

with a an average relative deviation lesser than 16%, over the entire range of Pe'_m and Sc.

(5) Dispersion out of Darcy domain (valid for $\text{Pe}_\text{m} > 10^6$) :

$$\frac{D_\text{L}}{D'_\text{m}} = \frac{\text{Pe}'_\text{m}}{2} \tag{4.72}$$

The correlations proposed are shown (see Fig. 4.20) to be significantly more accurate than previous correlations (see Fig. 4.18) and they cover the entire spectrum of values of Pe_m and Sc expected to be useful. It is important to have in mind that Eqs. 4.68–4.72 are recommended only for random packings of approximately "isometric" particles.

4.6.2 New Correlations: Radial Dispersion

For gas flow, by simply adding the contributions of molecular diffusion and turbulent dispersion, we suggested the well-known equation (in dimensionless form)

$$\frac{D_T}{D'_m} = 1 + \frac{Pe'_m}{12} \tag{4.73}$$

with a an average relative deviation lesser than 12%. Equation 4.73 give the correct asymptotic behaviour (both for very high and very low Pe_m) for both gases and liquids, as reported by several workers (see [65, 144]).

In the intermediate range of Pe_m they are still a reasonable approximation for gases, the wider deviation being observed in the intervals $3 < Pe_m < 300$, as shown in Fig. 4.21.

For the case of liquid flow in a porous media, our group used the division in four dispersion regimes to obtain the expressions presented below.

(1) Diffusion regime (valid for $Pe_m < 1$) :

$$\frac{D_T}{D'_m} = 1 \tag{4.74}$$

(2) Predominant mechanical dispersion (valid for $1 < Pe_m < 1600$) :

$$\frac{D_T}{D'_m} = 1 + \frac{1}{2.7 \times 10^{-5} Sc + 12/Pe'_m} \quad \text{for } Sc < 550 \tag{4.75a}$$

$$\frac{D_T}{D'_m} = 1 + \frac{1}{0.017 + 14/Pe'_m} \quad \text{for } Sc \geq 550 \tag{4.75b}$$

with a deviation lesser than 8 and 5%, respectively, over the entire range of Pe'_m and Sc.

(3) Pure mechanical dispersion (valid for $1,600 < Pe_m < 10^6$):

$$\frac{D_T}{D'_m} = \frac{Pe'_m}{(0.058Sc + 14) - (0.058Sc + 2)\exp\left(-\frac{500Sc^{0.5}}{Pe'_m}\right)} \quad \text{for } Sc < 550 \tag{4.76a}$$

$$\frac{D_T}{D'_m} = \frac{Pe'_m}{45.9 - 33.9 \times \exp\left(-\frac{21Sc}{Pe'_m}\right)} \quad \text{for } Sc \geq 550 \tag{4.76b}$$

and the experimental data do not deviate by more than 6 and 4% from the values given by Eqs. 4.76a and 4.76b, respectively.

(4) Dispersion out of Darcy domain (valid for $Pe_m > 10^6$):

$$\frac{D_T}{D'_m} = \frac{Pe'_m}{12} \tag{4.77}$$

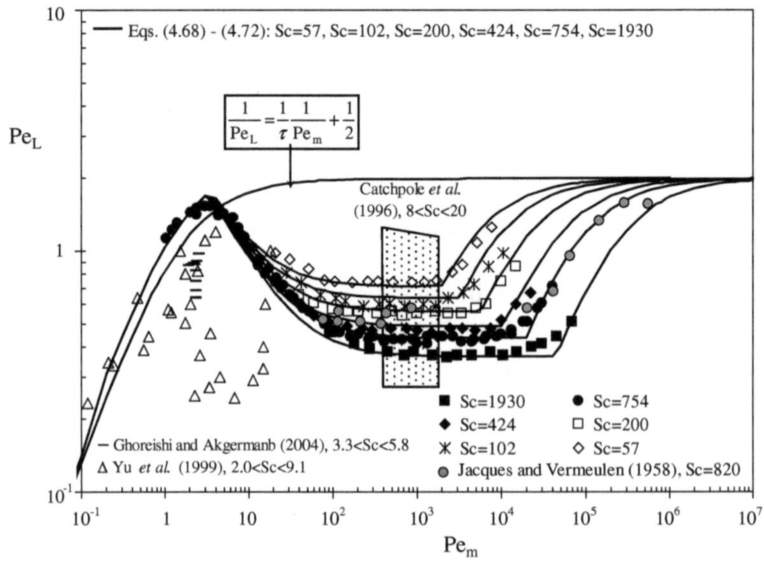

Fig. 4.20 Comparison between experimental data and correlations given by Eqs. 4.68–4.72

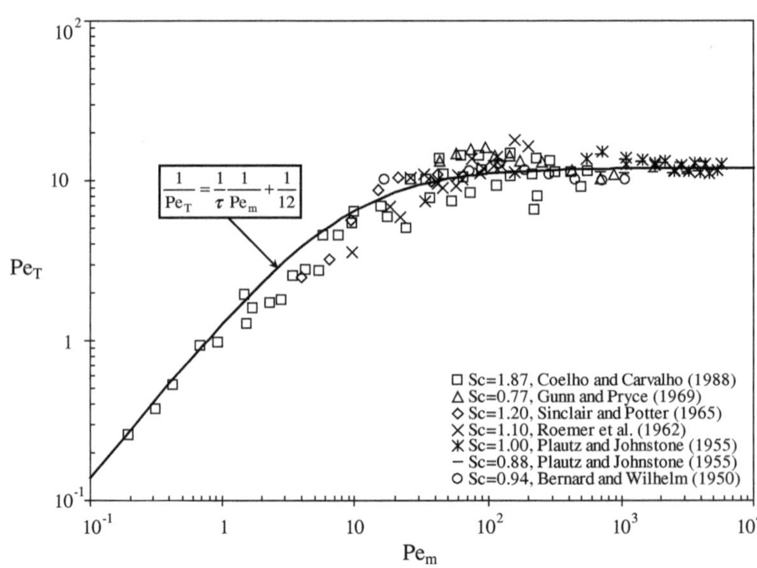

Fig. 4.21 Radial dispersion in gas flow

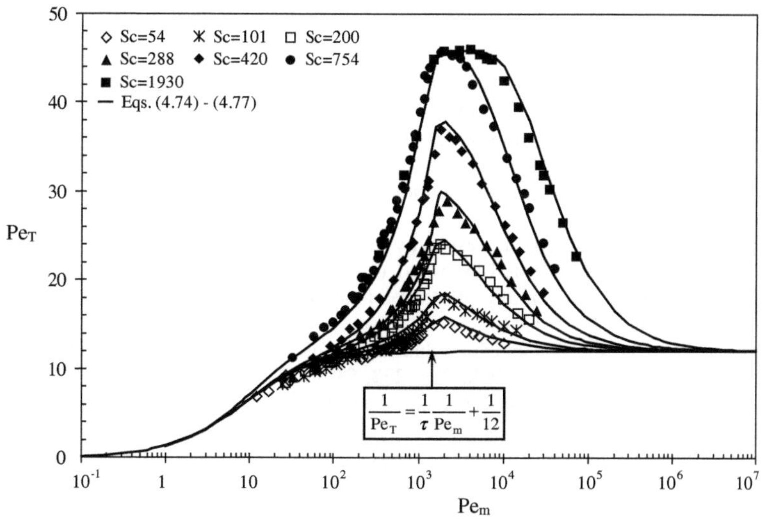

Fig. 4.22 Comparison between experimental data and our correlations, for radial dispersion

The experimental data are shown in Fig. 4.22, alongside the solid lines corresponding to Eqs. 4.74–4.77, for the values of Sc indicated in the figure. The agreement is seen to be generally very good, even when the values of Pe_T are represented on a linear scale. For Sc > 550 (see experiments with values of Sc of 754 and 1,930), the above equations representing the data must take into account that Pe_T is only dependent on Pe_m, in the ascending part of the curve Pe_T *vs.* Pe_m and that Pe_T only depends on $Re(= \varepsilon\, Pe_m/Sc)$, in the descending part of the same curve.

In conclusion, the present chapter increases our knowledge about diffusion and dispersion in packed beds by providing a critical analysis on the effect of fluid properties and porous medium on the values of axial and radial dispersion coefficients.

Different experimental techniques are presented in full detail and the data obtained from these techniques are very similar. An improved technique for the determination of the coefficient of radial dispersion in fluid flow through packed beds is described more detailed, which is based on the measurement of the rate of dissolution of buried flat or cylindrical surfaces.

A large number of experimental data on dispersion available in literature for packed beds were examined to pave the way for the formulation of new correlations for the prediction of Pe_T and Pe_L. The correlations proposed are shown to be more accurate than previous correlations and they cover the entire range of values of Pe_m and Sc. The axial dispersion coefficient can be calculated by Eqs. 4.68–4.72 and the radial dispersion coefficient by Eqs. 4.73–4.77.

References

1. Adler, P.M., Jacquin, C.G., Thovert, J.F.: The formation factor of reconstructed porous media. Water Resour. Res. **28**, 1571–1576 (1992)
2. Ahn, B.J., Zoulalian, A., Smith, J.M.: Axial dispersion in packed beds with large wall effect. AIChE J. **32**, 170–174 (1986)
3. Akehata, T., Sato, K.: Flow distribution in packed beds. Chem. Eng. Jpn. **22**, 430–436 (1958)
4. Archie, G.E.: The electrical resistivity log as an aid in determining some reservoir characteristics. Trans. AIME **146**, 54–61 (1942)
5. Aris, R.: On the dispersion of a solute in a fluid flowing through a tube. Proc. Roy. Soc. A, Math. Phy. **235**, 67–77 (1956)
6. Aris, R.: On the dispersion of a solute by diffusion, convection and exchange between phases. Proc. Roy. Soc. A, Math. Phy. **252**, 538–550 (1959)
7. Baron, T.: Generalized graphical method for the design of fixed bed catalytic reactors. Chem. Eng. Prog. **48**, 118–124 (1952)
8. Baumeister, E., Klose, U., Albert, K., Bayer, E.: Determination of the apparent transverse and axial dispersion coefficients in a chromatographic column by pulsed field gradient nuclear magnetic resonance. J. Chromatogr. A **694**, 321–331 (1995)
9. Bear, J.: Dynamics of Fluids in Porous Media. Dover Publications, New York (1972)
10. Bear, J., Verruijt, A.: Modelling Groundwater Flow and Pollution, 1st edn. Springer, Berlin (1987)
11. Beck, R.E., Schultz, J.S.: Hindered diffusion in microporous membranes with known pore geometry. Sci **170**, 1302–1305 (1970)
12. Benneker, A.H., Kronberg, A.E., Post, J.W., Van der Ham, A.G.J., Westerterp, K.R.: Axial dispersion in gases flowing though a packed bed at elevated pressures. Chem. Eng. Sci. **51**, 2099–2108 (1996)
13. Beran, M.J.: Dispersion of soluble matter in slowly moving fluids. Dissertation, Harvard University, Cambridge (1955)
14. Bernard, R.A., Wilhelm, R.H.: Turbulent diffusion in fixed beds of packed solids. Chem. Eng. Prog. **46**, 233–244 (1950)
15. Bischoff, K.B.: A note on gas dispersion in packed beds. Chem. Eng. Sci. **24**, 607–608 (1969)
16. Blackwell, R.J.: Laboratory studies of microscopic dispersion phenomena. Soc. Petrol. Eng. J. **225**, 1–8 (1962)
17. Blackwell, R.J., Rayne, J.R., Terry, W.M.: Factors influencing the efficiency of miscible displacement. Petrol. Trans. AIME **216**, 1–8 (1959)
18. Boudreau, B.P.: The diffusive tortuosity of fine-grained unlithified sediments. Geochim. Cosmochim. Acta **60**, 3139–3142 (1996)
19. Brenner, H.: The diffusion model of longitudinal mixing in beds of finite length. Numerical values. Chem. Eng. Sci. **17**, 229–243 (1962)
20. Brenner, H.: Dispersion resulting from flow through spatially periodic porous media. Philos. Trans. R. Soc. A **297**, 81–133 (1980)
21. Bruggeman, D.A.G.: Calculation of different physical constants of heterogeneous substances i, dielectric constant and conductivity of media of isotropic substances. Ann. Phys. Ser.-5 **24**, 636–664 (1935)
22. Bruinzeel, C., Reman, G.H., van der Laan, E.T.H.: Axial dispersion. Proceedings of the 3rd Congress of the European Federation of Chemical Engineering, Olympia, London, UK (1962)
23. Cairns, E.J., Prausnitz, J.M.: Longitudinal mixing in packed beds. Chem. Eng. Sci. **12**, 20–34 (1960)
24. Carberry, J.J.: Chemical and Catalytic Reaction Engineering. McGraw-Hill Chemical Engineering Series, USA (1976)

25. Carberry, J.J., Bretton, R.H.: Axial dispersion of mass in flow through fixed beds. AIChE J. **4**, 367–375 (1958)
26. Carbonell, R.G., Whitaker, S.: Dispersion in pulsed systems. Part II: Theoretical developments for passive dispersion in porous media. Chem. Eng. Sci. **38**, 1795–1801 (1983)
27. Catchpole, O.J., Berning, R., King, M.B.: Measurement and correlations of packed bed axial dispersion coefficients in supercritical carbon dioxide. Ind. Eng. Chem. Res. **35**, 824–828 (1996)
28. Chantong, A., Massoth, F.E.: Restrictive diffusion in aluminas. AIChE J. **29**, 725–731 (1983)
29. Chao, R., Hoelscher, H.E.: Simultaneous axial dispersion and adsorption in packed beds. AIChE J. **12**, 271–278 (1966)
30. Choudhary, M., Szekely, J., Weller, S.W.: The effect of flow maldistribution on conversion in a catalytic packed bed reactor. AIChE J. **22**, 1021–1032 (1976)
31. Chung, S.F., Wen, C.Y.: Longitudinal dispersion of liquid flowing through fixed and fluidized beds. AIChE J. **14**, 857–866 (1968)
32. Coelho, D., Thovert, J.F., Adler, P.M.: Geometrical and transport properties of random packings of spheres and aspherical particles. Phys. Rev. E **55**, 1959–1978 (1997)
33. Coelho, M.A.N., Guedes de Carvalho, J.R.F.: Transverse dispersion in granular beds. Part I: Mass transfer from a wall and the dispersion coefficient in packed beds. Chem. Eng. Res. Des. **66**, 165–177 (1988)
34. Collins, M.: Velocity distribution in packed beds. Dissertation, University of Delaware (1958)
35. Comiti, J., Renaud, M.: A new model for determining mean structure parameters of fixed beds from pressure drops measurements, application to beds packed with parallelepipedal particles. Chem. Eng. Sci. **44**, 1539–1545 (1989)
36. Crank, J.: The Mathematics of Diffusion, 2nd edn. Oxford University Press, Oxford (1975)
37. Danckwerts, P.V.: Continuous flow systems. Chem. Eng. Sci. **2**, 1–13 (1953)
38. De Jong, G.J.: Longitudinal and transverse diffusion in granular deposits. Trans. AGU **39**, 67–75 (1958)
39. Deisler, P.F., Wilhelm, R.H.: Diffusion in beds of porous solids - measurement by frequency response techniques. Ind. Eng. Chem. **45**, 1219–1227 (1953)
40. Delgado, J.M.P.Q., Guedes de Carvalho, J.R.F.: Measurement of the coefficient of transverse dispersion in packed beds over a range of values of Schmidt number 50–1000. Transp. Porous Med. **44**, 165–180 (2001)
41. Delmas, H., Froment, G.F.: Simulation model accounting for structural radial nonuniformities in fixed bed reactors. Chem. Eng. Sci. **43**, 2281–2287 (1988)
42. Dorweiler, V.P., Fahien, R.W.: Mass transfer at low flow rates in a packed column. AIChE J. **5**, 139–144 (1959)
43. Dullien, F.A.L.: Single phase flow through porous media and pore structure. Chem. Eng. J. **10**, 1–34 (1975)
44. Dullien, F.A.L.: Porous media, fluid transport and pore structure. Academic, San Diego (1979)
45. Ebach, E.A., White, R.R.: Mixing of fluids flowing through beds of oacked solids. AIChE J. **4**, 161–169 (1958)
46. Edwards, M.F., Helail, T.R.: Axial dispersion in porous media. Proceedings of the 2nd European Conference of Mixing, BHRA Fluids Engineering, Cranfield, UK (1977)
47. Edwards, M.F., Richardson, J.F.: Gas dispersion in packed beds. Chem. Eng. Sci. **23**, 109–123 (1968)
48. Eidsath, A., Carbonell, R.G., Whitaker, S., Herrmann, L.R.: Dispersion in pulsed systems. Part III: Comparison between theory and experiments for packed beds. Chem. Eng. Sci. **38**, 1803–1816 (1983)
49. England, R., Gunn, D.J.: Dispersion, pressure drop, and chemical reaction in packed beds of cylindrical particles. Trans. IChemE. **48**, T265–T275 (1970)

50. Evans, E.V., Kenney, C.N.: Gaseous dispersion in packed beds at low Reynolds numbers. Trans. IChemE. **44**, T189–T197 (1966)
51. Fahien, R.W., Smith, J.M.: Mass transfer in packed beds. AIChE J. **1**, 28–37 (1955)
52. Fetter, C.W.: Contaminant hydrogeology, 2nd edn. Prentice-Hall, New Jersey (1999)
53. Freeze, R.A., Cherry, J.A.: Groundwater. Prentice-Hall Inc, Englewood Cliffs (1979)
54. Fried, J.J., Combarnous, M.A.: Dispersion in porous media. Advances in Hydroscience n° 7. Academic, New York (1971)
55. Froment, G.F., Bischoff, K.B.: Chemical reactor analysis and design, 2nd edn. Wiley, New York (1990)
56. Ghoreishi S.M., Akgermanb, A.: Dispersion coefficients of supercritical fluid in fixed beds. Sep. Purif. Technol. **39**, 39–50 (2004)
57. Gibbs, S.J., Lightfoot, E.N., Root, T.W.: Protein diffusion in porous gel filtration chromatography media studied by pulsed field gradient NMR spectroscopy. J. Phys. Chem. **96**, 7458–7462 (1992)
58. Grane, F.E., Gardner, G.H.F.: Measurements of transverse dispersion in granular media. J. Chem. Eng. Data **6**, 283–287 (1961)
59. Gray, W.G.: A derivation of the equations for multi-phase transport. Chem. Eng. Sci. **30**, 229–233 (1975)
60. Greenkorn, R.A., Kessler, D.P.: Dispersion in heterogeneous nonuniform anisotropic porous media. Ind. Eng. Chem. **61**, 8–15 (1969)
61. Guedes de Carvalho, J.R.F., Delgado, J.M.P.Q.: The effect of fluid properties on dispersion in flow through packed. AIChE J. **49**, 1980–1985 (2003)
62. Guedes de Carvalho, J.R.F., Delgado, J.M.P.Q.: Radial dispersion in liquid flow through packed beds for $50 < Sc < 750$ and $103 < Pem < 105$. Proceedings of the 5th World Conference on Experimental Heat Transfer, Fluid Mechanics and Thermodynamics, Thessaloniki, Greece (2001)
63. Guedes de Carvalho, J.R.F., Delgado, J.M.P.Q.: Lateral dispersion in liquid flow through packed beds at $Pe_m < 1{,}400$. AIChE J. **46**, 1089–1095 (2000)
64. Gunn, D.J.: Axial and radial dispersion in fixed beds. Chem. Eng. Sci. **42**, 363–373 (1987)
65. Gunn, D.J.: Theory of axial and radial dispersion in packed beds. Trans. IChemE. **47**, T351–T359 (1969)
66. Gunn, D.J.: Mixing in packed and fluidised beds. Chem. Eng. J. **5**, E153–CE172 (1968)
67. Gunn, D.J., Malik, A.A.: Flow through expanded beds of solids. Trans. IChemE. **44**, T371–T379 (1966)
68. Gunn, D.J., Pryce, C.: Dispersion in packed beds. Trans. IChemE. **47**, T341–T350 (1969)
69. Han, N.W., Bhakta, J., Carbonell, R.G.: Longitudinal and lateral dispersion in packed beds, effect of column length and particle size distribution. AIChE J. **31**, 277–288 (1985)
70. Haring, R.E., Greenkorn, R.A.: Statistical model of a porous medium with nonuniform pores. AIChE J. **16**, 477–483 (1970)
71. Harleman, D.R.F., Rumer, R.: Longitudinal and lateral dispersion in an isotropic porous medium. J. Fluid Mech. **16**, 1–12 (1963)
72. Hartman, M.E., Wevers, C.J.H., Kramers, H.: Lateral diffusion with liquid flow through a packed bed of ion-exchange particles. Chem. Eng. Sci. **9**, 80–82 (1958)
73. Hassel, H.L., Bondi, A.: Mixing of viscous non-newtonian fluids in packed beds. AIChE J. **11**, 217–221 (1965)
74. Hennico, A., Jacques, G., Vermeulen, T.: Longitudinal dispersion in single-phase liquid flow through ordered and random packings. Lawrence Rad. Laboratory Report UCRL, 10696 (1963)
75. Hiby, J.W.: Longitudinal and transverse mixing during single-phase flow through granular beds. In: Interaction between fluid and particles, London IChemE, pp. 312–325, (1962)
76. Hiby, J.W., Schummer, P.: Zur messung der transversalen effektiven diffusion in durchstromten fullkorpersaulen. Chem. Eng. Sci. **13**, 69–74 (1960)
77. Hilal, M., Brunjail, D., Combi, J.: Electrodiffusion characterization of non-newtonian flow through packed-beds. J. Appl. Electrochem. **21**, 1087–1090 (1991)

78. Hoopes, J.A., Harleman, D.R.F.: Waste water recharge and dispersion in porous media. MIT Hydrodynamics Laboratory Report **75**, 55–60 (1965)
79. Jacques, G.L., Vermeulen, T.: Longitudinal dispersion in solvent-extraction columns, peclet numbers for random and ordered packings. University of California Rad. Laboratory Report N° 8029, US Atomic Energy Commission, Washington DC, USA (1958)
80. Johnson, G.W., Kapner, R.S.: The dependence of axial-dispersion on non-uniform flows in beds of uniform packing. Chem. Eng. Sci. **45**, 329–3339 (1990)
81. Klinkenberg, A., Krajenbrink, H.J., Lauwerier, H.A.: Diffusion in a fluid moving at uniform velocity in a tube. Ind. Eng. Chem. **45**, 1202–1208 (1953)
82. Klotz, D.: Untersuchungen zur dispersion in porosen medien. Z. Deutsch. Geol. Ges. **124**, 523–533 (1973)
83. Koch, D.C., Brady, J.F.: Dispersion in fixed beds. J. Fluid Mech. **154**, 399–427 (1985)
84. Koplik, J., Redner, S., Wilkinson, D.: Transport and dispersion in random networks with percolation disorder. Phys. Rev. A **37**, 2619–2636 (1988)
85. Kramers, H., Alberda, G.: Frequency response analysis of continuous flow systems. Chem. Eng. Sci. **2**, 173–181 (1953)
86. Langer, G., Roethe, A., Roethe, K.P., Gelbin, D.: Heat and mass-transfer in packed beds. Part III: Axial mass dispersion. Int. J. Heat Mass Tran. **21**, 751–759 (1978)
87. Latinen, G.A.: Mechanism of fluid-phase mixing in fixed and fluidised beds of uniformly sized spherical particles. Dissertation, Princeton University (1951)
88. Lerou, J.J., Froment, G.F.: Velocity, temperature and conversion profiles in fixed bed catalytic reactors. Chem. Eng. Sci. **32**, 853–861 (1977)
89. Levenspiel, O., Smith, W.K.: Notes on the diffusion-type model for the longitudinal mixing of fluids in flow. Chem. Eng. Sci. **6**, 227–233 (1957)
90. Liles, A.W., Geankopolis, C.J.: Axial diffusion of liquids in packed beds and end effects. AIChE J. **6**, 591–595 (1960)
91. List, E.J.: The stability and mixing of a density-stratified horizontal flow in a saturated porous medium. WM Keck Lab Hydraulics Water Resources Reporter KH-R-11, USA (1965)
92. Marsily, G.: Quantitative Hydrogeology, 1st edn. Academic, Orlando (1986)
93. Maxwell, J.C.: Treatise on Electricity and Magnetism. Oxford University Press, London (1873)
94. McHenry, J.R., Wilhelm, R.H.: Axial mixing of binary gas mixtures flowing in a random bed of spheres. AIChE J. **3**, 83–91 (1957)
95. Miller, S.T., King, C.J.: Axial dispersion in liquid flow through packed beds. AIChE J. **12**, 767–773 (1966)
96. Millington, R.J., Quirk, J.P.: Transport in porous media. Proceedings of the 7th International Congress of Soil Science, Madison, Wisc., USA (1960)
97. Miyauchi, T., Kikuchi, T.: Axial dispersion in packed beds. Chem. Eng. Sci. **30**, 343–348 (1975)
98. Niemann, E.H.; Dispersion during flow nonuniform heterogeneous porous media. Dissertation, Purdue University (1969)
99. Ogata, A.: The spread of a dye stream in an isotropic granular medium. US Geol. Surv. Prof. Pap. 411-G, pp. 1–11. USA (1964)
100. Otake, T., Kunugita, E.: Axial dispersion of the gas phase in countercurrent packed bed columns. Chem. Eng. Jpn. **22**, 144–150 (1958)
101. Payne, L.W., Parker, H.W.: Axial dispersion of non-newtonian fluids in porous media. AIChE J. **19**, 202–204 (1973)
102. Perkins, T.K., Johnston, O.C.: A review of diffusion and dispersion in porous media. Soc. Petrol. Eng. J. **2**, 70–84 (1963)
103. Pfannkuch, H.O.: Contribution a l'etude des déplacements de fluids miscibles dans un milieu poreux. Rev. Inst. Franc. Pétrol. **18**, 215–219 (1963)
104. Pickens, J.F., Grisak, G.E.: Scale-dependent dispersion in a stratified granular aquifer. Water Resour. Res. **17**, 1191–1211 (1981)

105. Plautz, D.A., Johnstone, H.F.: Heat and mass transfer in packed beds. AIChE J. **1**, 193–199 (1955)
106. Pozzi, A.L., Blackwell, R.J.: Design of laboratory models for study of miscible displacement. Soc. Petrol. Eng. J. **3**, 28–40 (1963)
107. Probst, K., Wohlfahrt, K.: Empirische abschätzung effektiver diffusionskoeffizienten in porösen systemen. Chem. Eng. Tech. **1**, 737–739 (1979)
108. Raimondi, P., Gardner, G.H.F., Petrick, C.B.: Effect of pore structure and molecular diffusion on the mixing of miscible liquids flowing in porous media. In: AIChE-SPE joint symposium, San Francisco (1959)
109. Rifai, M.N.E., Kaufman, W.J., Todd, D.K.: Dispersion phenomena in laminar flow through porous media. University of California, Sanitary Engineering Report 3. Inst. Eng. Res. Ser. **90**, 1–157 (1956)
110. Roblee, L.H.S., Baird, R.M., Tierney, J.W.: Radial porosity variations in packed beds. AIChE J. **4**, 460–468 (1958)
111. Roemer, G., Dranoff, J.S., Smith, J.M.: Diffusion in packed beds at low flow rates. Ind. Eng. Chem. Fund. **1**, 284–287 (1962)
112. Rosenberg, D.V.: Axial dispersion in porous media. J. Am. Inst. Chem. Engrs. **2**, 55–64 (1956)
113. Saffman, P.G.: A theory of dispersion in a porous medium. J. Fluid Mech. **6**, 321–325 (1959)
114. Saffman, P.C.: Dispersion in flow through a network of capillaries. J. Fluid Mech. **7**, 194–207 (1960)
115. Sahimi, M.: Flow and transport in porous media and fractured rock. VCH Velagsgesellschaft mbH, Weinheim (1995)
116. Scheidegger, A.E.: The physics of flow through porous media, 3rd edn. University of Toronto Press, Toronto (1974)
117. Schuster, J., Vortmeyer, D.: Ein einfaches verfahren zur näherungsweisen bestimmung der porosität in schttungen als funktion des wasdabstandes. Chem. Eng. Tech. **52**, 848–855 (1980)
118. Schwartz, C.E., Smith, J.M.: Flow distribution in packed beds. Ind. Eng. Chem. **45**, 1209–1218 (1953)
119. Scott, D.S., Lee, W., Papa, J.: The measurement of transport coefficients in gas-solid heterogeneous reactions. Chem. Eng. Sci. **29**, 2155–2167 (1974)
120. Sen, P.N., Scala, C., Cohen, M.H.: A self-similar model for sedimentary-rocks with application to the dielectric-constant of fused glass-beads. Geophysics **46**, 781–795 (1981)
121. Sherwood, T.K., Pigford, R.L., Wilke, C.R.: Mass transfer. McGraw-Hill, Tokyo (1975)
122. Shimamura, K.: Gas diffusion through compacted sands. Soil Sci. **153**, 274–279 (1992)
123. Simpson, E.S.: Transverse dispersion in liquid flow through porous media. US Geol. Surv. Prof. Pap. 411-C, pp. 1–30. USA (1962)
124. Sinclair, R.J., Potter, O.E.: The dispersion of gas in flow through a bed of packed solids. Trans. IChemE. **43**, T3–T9 (1965)
125. Singer, E., Wilhelm, R.H.: Heat transfer in packed beds, analytical solution and design method. Fluid flow, solids flow and chemical reaction. Chem. Eng. Prog. **46**, 343–352 (1950)
126. Slattery, J.C.: Momentum, energy and non transfer in continua. McGraw-Hill, New York (1972)
127. Slichter, C.S.: Field measurement of the rate of movement of underground waters. US Geol. Surv. Water Supply Pap. 140, USA (1905)
128. Smith, W.D., Bretton, R.H.: Paper presented at AIChE Houston meeting, Texas, USA (1967)
129. Stephenson, J.L., Stewart, W.E.: Optical measurements of porosity and fluid motion in packed beds. Chem. Eng. Sci. **41**, 2161–2170 (1986)
130. Strang, D.A., Geankopolis, C.J.: Longitudinal diffusivity of liquids in packed beds. Ind. Eng. Chem. **50**, 1305–1308 (1958)

131. Suzuki, M., Smith, J.M.: Dynamics of diffusion and adsorption in a single catalyst pellet. AIChE J. **18**, 326–333 (1972)
132. Tan, C.S., Liou, D.C.: Axial dispersion of supercritical carbon dioxide in packed beds. Ind. Eng. Chem. Res. **28**, 1246–1250 (1989)
133. Taylor, G.: Dispersion of soluble matter in solvent flowing slowly through a tube. Proc. Roy. Soc. A, Math. Phy. **219**, 186–203 (1953)
134. Towle, W.L., Sherwood, T.K.: Studies in eddy diffusion. Ind. Eng. Chem. **31**, 457–467 (1939)
135. Tsotsas, E., Schlunder, E.U.: On axial dispersion in packed beds with fluid flow. Chem. Eng. Proc. **24**, 15–31 (1988)
136. Urban, J.C., Gomezplata, A.: Axial dispersion coefficients in packed beds at low Reynolds numbers. Can. J. Chem. Eng. **47**, 353–363 (1969)
137. Vortmeyer, D., Schuster, J.: Evaluation of steady flow profiles in rectangular and circular packed beds by a variational method. Chem. Eng. Sci. **38**, 1691–1699 (1983)
138. Vortmeyer, D., Winter, R.P.: Impact of porosity and velocity distribution of the theoretical prediction of fixed-bed chemical reactor performance, comparison with experimental data. ACS symposium series, pp. 49–61. USA (1982)
139. Weissberg, H.L.: Effective diffusion coefficient in porous media. J. Appl. Phys. **34**, 2636–2639 (1963)
140. Wen, C.Y., Fan, L.S.: Axial dispersion of non-Newtonian liquids in fluidized beds. Chem. Eng. Sci. **28**, 1768–1773 (1973)
141. Wen, C.Y., Fan, L.T.: Models for systems and chemical reactors. Marcel Dekker, New York (1975)
142. Wen, C.Y., Yin, J.: Axial dispersion of a non-newtonian liquid in a packed bed. AIChE J. **17**, 1503–1504 (1971)
143. Whitaker, S.: Diffusion and dispersion in porous media. AIChE J. **13**, 420–432 (1967)
144. Wilhelm, R.H.: Progress towards the a priori design of chemical reactors. Pure Appl. Chem. **5**, 403–421 (1962)
145. Wilson, H.A.: On convection of heat. P. Camb. Philos. Soc. **12**, 406–423 (1904)
146. Wronski, S., Molga, E.: Axial dispersion in packed beds: the effect of particle size non-uniformities. Chem. Eng. Prog. **22**, 123–135 (1987)
147. Yu, D., Jackson, K., Harmon, T.C.: Dispersion and diffusion in porous media under supercritical conditions. Chem. Eng. Sci. **54**, 357–367 (1999)
148. Yun, M.J., Yu, B.M., Zhang, B., Huang, M.Y.: A geometry model for tortuosity of streamtubes in porous media with spherical particles. Chin. Phys. Lett. **22**, 1464–1467 (2005)

Chapter 5
Modeling of Transport Processes in Porous Materials

5.1 Introduction

As far as porous materials are characterized by high complexity in their *invisible internal structure*, modeling seems to be a grateful tool for understanding the transport processes that take place in porous media. Several modeling approaches, applicable in porous domain, have been introduced and used in a variety of scientific and industrial applications. Numerous industrial and technological applications involving fluid flow and mass transport processes within multi-particle assemblages have attracted scientific interest in recent decades. These focus mainly on industrial physicochemical processes (e.g. sedimentation, catalysis), alternative energy sources (e.g. fuel cells) and separation techniques (e.g. chromatography, filters). This chapter focuses on the modeling of mass transport in granular structures as well as on the estimation of macroscopic quantities (such as mass transport coefficient, adsorption efficiency, etc.).

5.2 Single Phase Transport in Unit Cells

Although arrays of regularly spatially distributed grains represent an idealization of real granular media, they have been widely studied from both the fluid dynamics and mass transport points of view. Due to their complex geometry, random particle distributions were the subject of few investigations until about twenty years ago. Since then, fast advances in computational capabilities have contributed to reviving research on this topic with emphasis on hydrodynamic aspects. Very often the particles are assumed to be sufficiently small and the physical properties of the fluid (i.e. viscosity and density) independent of the concentration of the transferred species, thus making it possible to uncouple the fluid problem from the mass-transport problem. Furthermore, the case of low Reynolds number flow is very important from an application point of view, where analytical solutions can be obtained for the flow field and the mass diffusion and/or adsorption process.

F. A. Coutelieris and J. M. P. Q. Delgado, *Transport Processes in Porous Media*, Advanced Structured Materials 20, DOI: 10.1007/978-3-642-27910-2_5,

5.2.1 Fundamental Quantities

Generally speaking, the transient mass transport of a passive solute in the void space of a porous medium is described by the convection–diffusion equation [4]:

$$\frac{\partial c_A}{\partial t} + \mathbf{v} \cdot \nabla c_A = D\nabla^2 c_A + R(c_A) \tag{5.1}$$

where c_A is the concentration of the passive solute, $\mathbf{v}$ is its velocity in the medium, D is its diffusivity in the intermediate occupying the medium's void space, and $R(c_A)$ describes the reactions taking place in the bulk phase.

For the sake of simplicity, the most common assumptions made in the relative literature and applied in the forthcoming analysis are:

- steady state condition (i.e. $\frac{\partial c_A}{\partial t} = 0$)
- inert bulk phase (i.e. $R(c_A) = 0$).

Consequently, the governing differential equation for mass transport in a porous material can often be written as:

$$\mathbf{v} \cdot \nabla c_A = D\nabla^2 c_A \tag{5.2}$$

The LHS of the above equation describes the mass transport *due to convection* while RHS stands for the mass transport *due to diffusion*. The relative dominance between these simultaneously occurring phenomena is expressed by the dimensionless *Peclet number*, defined as:

$$Pe = \frac{\text{characteristic velocity} \cdot \text{characteristic length}}{\text{diffusivity}} = \frac{U \cdot L}{D} \tag{5.3}$$

Obviously, the higher the Pe, the more convective the mass transport regime.

The microscopic quantity c_A is related to useful macroscopic quantities, such as the mass transport coefficient, the Sherwood (Nusselt) number, and the adsorption efficiency. A brief description of these quantities is given below.

The mass transport coefficient, k_0, describes the ability of the mass to be transferred from the bulk phase to solid absorbers. It is obviously dependant on the physical properties of the materials involved in the procedure, the porous structure, and some physical parameters such as pressure, temperature and volumetric flow-rate, which could be considered critical for the process. Mathematically, the mass transport coefficient is defined by the expression:

$$k_0(\Delta c_A)S_{abs} = -\iint_{S_{abs}} [-N_A]_{surf}\, dS_{abs} \tag{5.4}$$

where Δc_A is the absolute difference of the concentration upon the absorbing surface S_{abs} from the incoming concentration, and $[-N_A]_{surf}$ is the component of molar flux which is normal on the collectors surface. The dimensionless overall

Sherwood number is a direct consequence of the mass transport coefficient and is defined as:

$$Sh_o = \frac{\text{mass transport coefficient} \cdot \text{characteristic length}}{\text{diffusivity}} = \frac{k_o \cdot L}{D} \tag{5.5}$$

For practical applications, the most useful macroscopic quantity is the adsorption efficiency, λ_0, defined by the ratio of the solute adsorption rate divided by the rate of the upstream influx:

$$\lambda_0 = 1 - \frac{\iint_{S_{outlet}} c_A \mathbf{v} \cdot \underline{n} dS}{\iint_{S_{inlet}} c_A \mathbf{v} \cdot \underline{n} dS} \tag{5.6}$$

where $\underline{\boldsymbol{n}}$ is the vector normal to the surface. For cases regarding the adsorption efficiency of a grain-in-cell, the above expression is modified to:

$$\lambda_0 = \frac{\iint\limits_{S_{abs}} [-N_{A_r}]_{surf} dS_{abs}}{u_\infty c_{A,\infty} S_{inlet}} \tag{5.7}$$

where u_∞ is the approaching velocity, $c_{A,\infty}$ is the constant bulk concentration sufficiently far away from the collector surface, and S_{inlet} is the inlet surface that the solute faces to pass through the porous medium.

5.2.2 Adsorption Mechanisms

The majority of scientific and technological applications that deal with mass transport in porous materials, involve a *sorption mechanism*. Generally speaking, sorption includes adsorption and ion exchange, processes that involve the transfer and resulting equilibrium distribution of one or more solutes between a fluid phase and a solid surface. More precisely, adsorption generally involves the accumulation (or depletion) of solute molecules at an interface (gas–solid and liquid–solid interfaces), with solute distributed selectively between the fluid and solid phases. On the other hand, ion exchange occurs when positively (cations) or negatively (anions) charged ions from the fluid replace dissimilar ions of the same charge, being initially in the solid surface. Furthermore, the partitioning of a single solute between fluid and sorbed phases can be also included in the sorption mechanism.

Phase equilibrium between liquid and sorbed phases for one or many components in adsorption, or two or more species in ion exchange, is considered the single most important factor affecting process performance. This equilibrium is usually expressed by a boundary condition upon the liquid–solid interface, while it is of great importance to obtain a simple expression for the detailed description of the sorption mechanism. In relevant literature, several different cases for the adsorption of the solute upon the solid surface have been considered and used.

The first and simplest case corresponds to instantaneous adsorption upon the solid surface [17] and can be expressed as:

$$c_A = 0, \text{ on the solid–liquid interface} \tag{5.8}$$

Despite its mathematical simplicity, it is clear that the instantaneous adsorption approach pertains to a very limited range of applications, being a fairly rare physicochemical phenomenon. Thus, detailed models referring to adsorption following isotherms have also been adopted. The most common is that of Langmuir's isotherm, which can be formulated as [12]:

$$D\underline{n} \cdot \nabla c_A = \frac{k}{K} c_s, \text{ on the solid–liquid interface} \tag{5.9}$$

where c_s is the concentration of the sorpted material, K is a constant defined by the Langmuir isotherm:

$$\Theta_{eq} = \frac{Kc_b}{1 + Kc_b} \tag{5.10}$$

and k is a reaction rate defined from the relation

$$R(c_s) = kc_b(c_{mx} - c_s) \tag{5.11}$$

where $R(c_s)$ is the overall adsorption rate given as a function of the surface concentration c_s, c_b is the concentration of the diluted mass in the neighborhood of the solid surface, c_{mx} is the maximum concentration attained when the surface is completely covered by substance A, and Θ_{eq} is the ratio of the covered to the total surface, defined as:

$$\Theta_{eq} = \frac{c_s}{c_{mx}} \tag{5.12}$$

Another more detailed approach is the more complicated but realistic sorption mechanism, which involves adsorption-reaction-desorption [13]. More specifically, it can be considered that the solute diluted in the bulk phase is initially adsorbed by the solid surface where a heterogeneous reaction takes place, and its products, assumed to be inactive and of very low concentrations, are again desorbed in the bulk phase. The adsorption is assumed to occur due to vacant sites [2] that are normally distributed over the surface area while the whole process is determined by an overall rate according to thermodynamics. The above mechanism is often described by an expression of the form:

$$D\underline{n} \cdot \nabla c_A = R_n, \text{on the solid–liquid interface} \tag{5.13}$$

With R_n denoting the overall adsorption rate, given as:

$$R_n = k_s c_{AS}^n \tag{5.14}$$

where k_s is the rate constant of the heterogeneous reaction upon the surface, and the concentration of solute on the solid surface, c_{AS}, is calculated by solving the non linear equation:

$$k_s c_{AS}^n + \left[k_A^d + k_A^a c_A)_{surf} N\right] c_{AS} - k_A^a\, c_A)_{surf}\, \xi_m = 0 \tag{5.15}$$

Equation 5.15 correlates the hard-to-measure surface concentration of A, c_{AS}, with its concentration in the bulk phase very close to the solid surface $c_A)_{surf}$ by considering the balance for the active sites on the adsorbing surface. In the above equation, the terms $k_A^a$and k_A^ddenote the adsorption and desorption rate constants of solute, respectively, ξ_m is the concentration of the vacant sites of the solid surface, and N is Avogadro's number, while n denotes the order of the heterogeneous reaction upon the surface. In general, only the cases of $n = 0,\ n = 1$ and $n = 2$ are of practical importance, but zero order reactions are also used in limited applications [2].

Finally, it should be noted that some models consider neutral solid surface i.e. absolute absence of adsorption [11]. This case can be described by the following boundary condition upon the solid surface:

$$\underline{\boldsymbol{n}} \cdot \nabla c_A = 0, \text{ on the solid–liquid interface} \tag{5.16}$$

5.2.3 Mass Transport Through Spheres

As discussed in Sect. 2.2, sphere-in-cell geometry is a simple model that represents the actual complex geometry of the pore space in spherical particle assemblages and the approximation of the flow-field therein. Consider a typical sphere-in-cell model where the approaching fluid is a dilute solution of a substance A, which is moving towards the solid adsorbing surface. After taking into account the plane symmetry of the problem, the governing equation for the steady state mass transport in the fluid phase within the model can be written in spherical coordinates (r, θ) as:

$$u_r \frac{\partial c_A}{\partial r} + \frac{u_\theta}{r}\frac{\partial c_A}{\partial \theta} = D\left(\frac{\partial^2 c_A}{\partial r^2} + \frac{2}{r}\frac{\partial c_A}{\partial r} + \frac{1}{r^2}\frac{\partial^2 c_A}{\partial \theta^2} + \frac{\cot\theta}{r^2}\frac{\partial c_A}{\partial \theta}\right) \tag{5.17}$$

where u_r and u_θ are the $r-$ and $\theta-$velocity components, given either by Eqs. 3.12 and 3.13 or by Eqs. 3.19 and 3.20, where the set chosen depends on the model considered.

The axial symmetry of the mass transfer problem is expressed by the following boundary conditions:

$$\left.\frac{\partial c_A}{\partial \theta}\right]_{\theta=\pi} = 0, \quad a < r \leq \beta \tag{5.18}$$

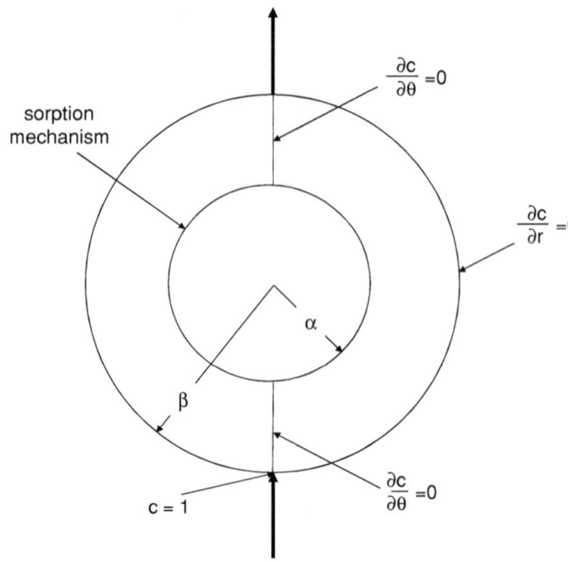

Fig. 5.1 Mass transport taking place in a sphere-in-cell. *Bold arrows* indicate flow direction

$$\left.\frac{\partial c_A}{\partial \theta}\right]_{\theta=0} = 0, \quad a < r \leq \beta \tag{5.19}$$

To ensure the continuity of concentration for any Peclet number, Coutelieris et al. [9] proposed the following boundary condition at the outer boundary of the cell:

$$c_A(r = \beta, \theta = \pi) = 1 \tag{5.20}$$

$$\left.\frac{\partial c_A}{\partial r}\right]_{r=\beta} = 0, \quad 0 \leq \theta < \pi \tag{5.21}$$

For high Peclet values, the boundary condition (5.20) and (5.21) is completely equivalent to the well-known Levich approach for unbounded fluids, given elsewhere as [8, 17]:

$$c_A = 1, \quad \text{for } r \to \infty \quad \text{or} \quad r = \beta \tag{5.22}$$

Figure 5.1 shows the extent of the concentration boundary layer surrounding the solid surface.

There are several approaches for describing the sorption mechanism (see Sect. 5.2.2). By using the geometrical specification of the sphere-in-cell model, sorption in specific spherical geometries can be written as follows:

$$[(5.16)\ \text{neutral surface}] \Rightarrow \left.\frac{\partial c_A}{\partial r}\right]_{r=\alpha} = 0, \quad 0 \leq \theta < \pi \tag{5.23}$$

$$[(5.8)\ \text{instantaneous adsorption}] \Rightarrow c_A(r = a, \theta) = 0,\ 0 \leq \theta \leq \pi \tag{5.24}$$

$$[(5.9)\text{ Langmuir isotherm}] \Rightarrow D\frac{\partial c_A}{\partial r}\bigg]_{r=\alpha} = \frac{k}{K}c_s, \quad 0 \le \theta \le \pi \tag{5.25}$$

$$[(5.13)\text{ adsorpionndash;reactionndash;desorption}] \Rightarrow D\frac{\partial c_A}{\partial r}\bigg]_{r=\alpha} = R_n,\ 0 \le \theta \le \pi \tag{5.26}$$

At high *Pe* numbers, the above described mass transport problem (considering each possible adsorption case) is amenable to analytical treatment with sufficient accuracy, as shown by various research groups [8, 17]. Mathematically, it is notable here that the RHS of Eq. 5.2 is simplified enough to accept a variables' separated function as an analytical solution. This is because the concentration boundary layer is very thin compared to the local radius of curvature of the particle, therefore the curvature term and the tangential diffusion terms can be neglected. In particular, Eq. 5.2 becomes:

$$u_r\frac{\partial c_A}{\partial r} + \frac{u_\theta}{r}\frac{\partial c_A}{\partial \theta} = D\frac{\partial^2 c_A}{\partial r^2} \tag{5.27}$$

which is parabolic on θ and can be solved analytically in a manner quite similar to that of Levich [17]. By using dimesionless quantities, the concentration profile in the fluid phase is of the form:

$$c_A(z) = c_2\int_0^z e^{-\frac{4}{9}t^3}\,dt + c_3 \tag{5.28}$$

where:

$$z = \sqrt[3]{\frac{3\,Pe}{4\alpha^3}}\,r\,\frac{\sin\theta}{\theta - 0.5\sin 2\theta} \tag{5.29}$$

and c_2 and c_3 are coefficients that can impose the appropriate boundary conditions for mass transport.

For the case of low and moderate Peclet numbers, all the terms of the transport equation (5.17) survive, therefore it is necessary to apply numerical approaches to obtain the solution. What is important for cases of low Peclet number is the boundary condition on the outer boundary, as described by Eqs. 5.20 and 5.21, is equivalent to that of Levich for high Peclet numbers, while it allows the boundary layer to violate the concept of a very thin boundary layer imposed by the Levich approach [17]. The difference between high and low Peclet mass transport in sphere-in-cell geometry, is graphically summarized in Fig. 5.2.

Obviously, mass transport problems assuming neutral solid grains (no adsorption) result in a uniform concentration profile without scientific or industrial interest. For instantaneous adsorption and using Kuwabara's approach for the velocity field, the concentration profiles for a moderate Peclet value (Pe = 70) are presented in Fig. 5.3.

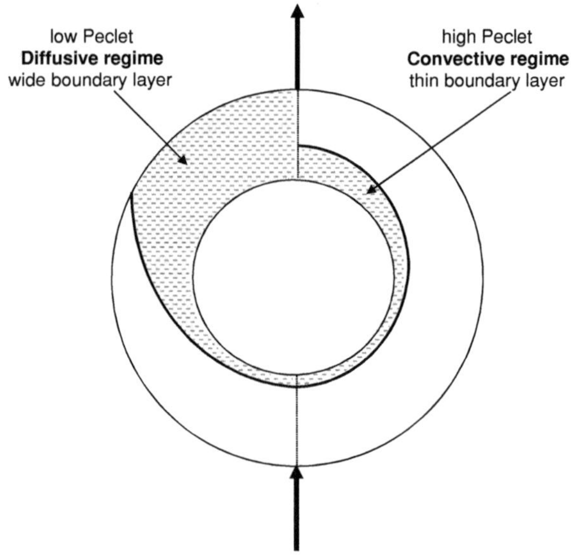

Fig. 5.2 Schematic visualization of boundary layers for low and high Peclet numbers

For Langmuir-type adsorption with the same geometry and flow conditions, the concentration profiles are shown in Fig. 5.4.

It should be stressed however, that the boundary condition (5.26) in combination with (5.14)–(5.15), imply non-linearity in the whole approach and, therefore, an iterative numerical technique is necessary to obtain the solution. Thus, the integration of the transport equation must be numerical, although the flow field can be obtained analytically through Happel or Kuwabara type boundary conditions. Figure 5.5 visualizes the concentration profiles for the realistic sorption mechanism with the same geometry and flow conditions as those for the other adsorption types, presented above.

The above Figs. 5.3, 5.4 and 5.5 show a gradual decrease of the concentration for constant r as the angular position approaches the outlet because of the shape and extent of the concentration boundary layer. Therefore, the possibility for the solute to be captured by the solid surface decreases with angular position because of the convective regime. In general, higher concentration gradients at any radial and angular positions are found in cases of instantaneous adsorption compared to those for other sorption mechanisms, as the surface concentration, $c_A(\alpha,\theta)$, is much higher in the cases of Langmuir type or realistic adsorption mechanisms, taking its highest value at the impact point and decreasing monotonically as θ tends to π.

For all sorption mechanisms considered, the overall Sherwood number, Sh_o, is calculated as follows [6]:

$$Sh_o = \frac{1}{4\pi a^2} \int_{\pi}^{0} \frac{\left(\frac{\partial c_A}{\partial r}\right)_{r=a}}{c_A(\beta,\theta) - c_A(\alpha,\theta)} d\theta \tag{5.30}$$

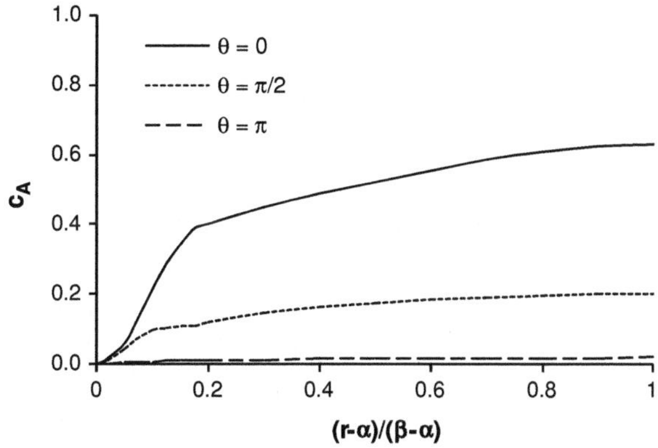

Fig. 5.3 Spatial concentration profiles for $\varepsilon = 0.9$ at three different angular positions of the sphere-in-cell for instantaneous adsorption

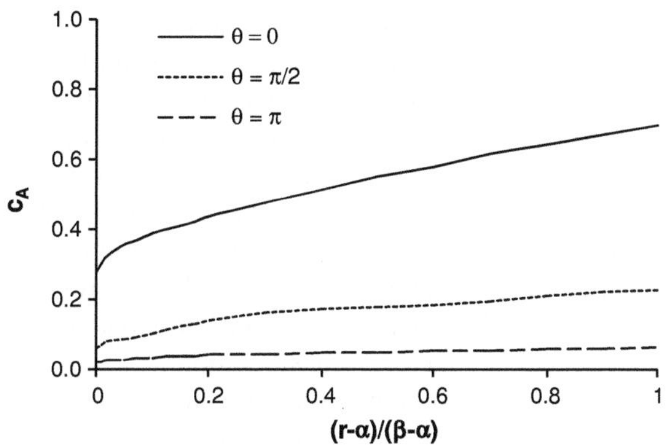

Fig. 5.4 Spatial concentration profiles for $\varepsilon = 0.9$ at three different angular positions of the sphere-in-cell for Langmuir type adsorption $\left(\frac{k}{K} = 1\right)$

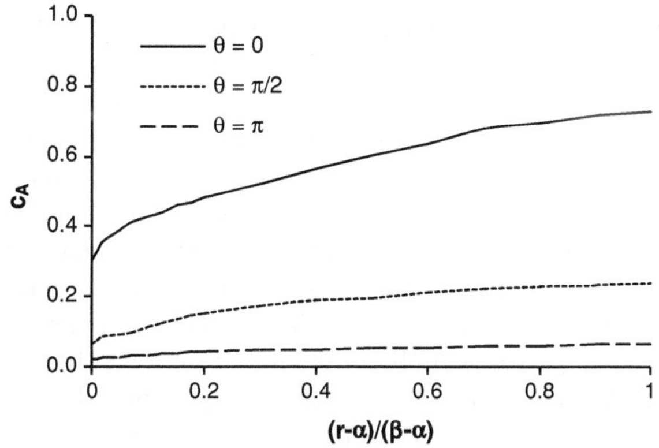

Fig. 5.5 Spatial concentration profiles for $\varepsilon = 0.9$ at three different angular positions of the sphere-in-cell for realistic sorption mechanism

Finally, the adsorption efficiency of a sphere-in-cell model can be written as:

$$\lambda_0 = \frac{\iint\limits_{S_{sphere}} [-N_{A_r}]_{r=a} dS_{sphere}}{u_\infty c_{A,\infty} 4\pi\beta^2} \tag{5.31}$$

where it can be easily shown for the specified spherical geometry that:

$$\lambda_o = \frac{a^2}{4\pi\beta^2} \int_0^\pi \sin\theta \left(\frac{\partial\left(\frac{c_A}{c_{A,\infty}}\right)}{\partial\left(\frac{r}{\alpha}\right)} \right)_{r=\alpha} d\theta \tag{5.32}$$

To initially investigate the macroscopic behavior of the sphere-in-cell model, Fig. 5.6 presents the overall Sherwood number as a function of the Peclet number for a relatively high porosity value ($\varepsilon = 0.9$) in the case of instantaneous adsorption. An almost monotonic increment of Sh_o is observed as the Peclet values increase. This is due to the higher concentration gradients close to the adsorbing surfaces in highly convective regimes. It has been also shown that such analytical results are in very good agreement with experimental measurements and numerical simulations [11]. This agreement can be attributed to the high porosity value used, taking into account that the analytical approach has been shown to be highly accurate in that range [5, 9, 26]. High porosity values favor the accuracy of transport results because mass transport under high Peclet values develops very thin boundary layers, a consideration which corresponds to an outer sphere of radius significantly higher than the inner one, i.e. porosity takes quite high values. To further clarify the situation, the dependence of the overall Sherwood number on porosity for the case of instantaneous adsorption is shown in Fig. 5.7. Evidently, increasing porosities do not favor mass exchange efficiency and the Sh_o drops. High porosity values correspond to low available surface for mass transfer leading to low adsorption rates for the porous material. Again experimental data and numerical predictions are generally in agreement with the analytical results [11] although an erratic behavior is observed. This behavior is due to the fact that high porosities, where analytical models are in principle applicable, correspond to a very low number of spheres for the numerical representation of the porous medium. It should be noted that at porosities lower than 0.8, the available analytical approaches become gradually less accurate and therefore less dependable.

To include the more detailed and realistic sorption mechanism of Eqs. 5.26, 5.14 and 5.15 into the whole approach, it is necessary to numerically solve the boundary value problem described above either using the Kuwabara or Happel expression for the velocity components. Frequently, a non-uniform finite-difference scheme is employed where the value of λ_0 can be calculated once the radial component of the concentration gradient upon the surface is known by using a numerical integration method (the most suitable choice could be the modified Newton–Cotes with an adjustable step). The effect of the sorption mechanism

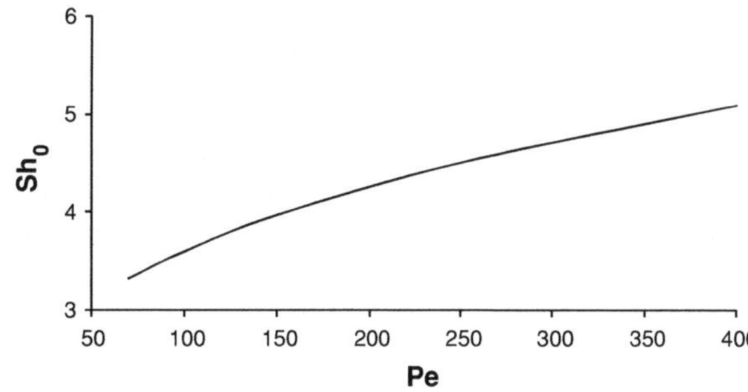

Fig. 5.6 Dependence of the overall Sherwood number on the Peclet for a high porosity value (instantaneous adsorption)

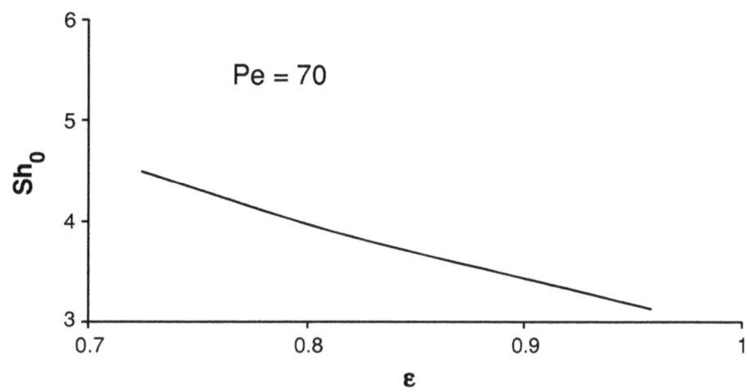

Fig. 5.7 Influence of the porosity on the overall Sherwood number (instantaneous adsorption)

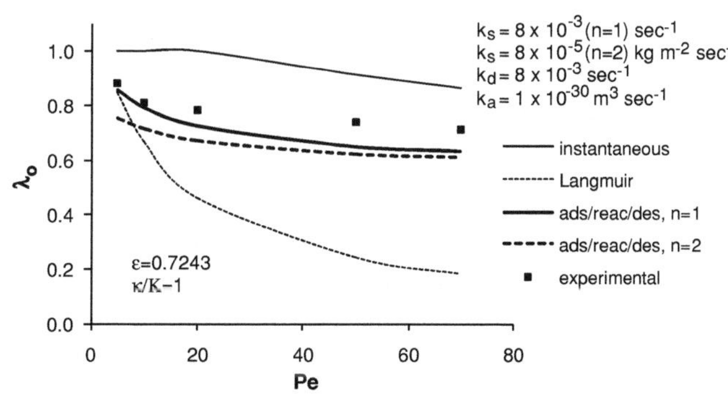

Fig. 5.8 Adsorption efficiency of the sphere-in-cell model as a function of Pe for several adsorption mechanisms, compared to experimental data presented by discrete points [29]

considered, is clearly depicted in Fig. 5.8, which presents the adsorption efficiency as a function of Pe for various adsorption mechanisms.

In general, higher efficiency is found for instantaneous adsorption than for the Langmuir type, as the concentration gradients are lower in the latter case. Indeed, in Langmuir adsorption concentration on the solid surface attains non zero values thus rendering the overall driving force $c_A(\beta, \theta) - c_A(a, \theta)$ smaller. A general decrement of the efficiency with Peclet is observed, a trend expected as the more convective flows (increasing Peclet) tend to prevent solute from being captured by the solid surface. The discrete points represent the experimental data of Wilson and Geankoplis [29] for the rather high porosity of 0.7 (the highest porosity used in their experiments). As these authors measured the overall Sherwood number, it is necessary to transform it into adsorption efficiency to allow direct comparison with the predictions. Starting from the definition of adsorption efficiency in Eq. 5.31 the numerator (r-component of the molar flux on the adsorbing surface) can be expressed by the use of the mass transfer coefficient as $k_0 \cdot \Delta C \cdot S_{sphere}$. After some algebraic manipulations, this leads to the following linear relation between the overall Sherwood number and λ_0:

$$Sh = \lambda_0 \frac{Pe}{(1-\varepsilon)} \frac{a}{L} \frac{c_{A.\infty}}{\langle c_A \rangle} \tag{5.33}$$

where $\langle c_A \rangle$ is the spatially averaged concentration of solute, and L is the characteristic length, used when defining the Peclet number (see Eq. 5.3) As the experimental adsorption mechanism is not clearly known, the agreement between predictions and experimental data is considered sufficient.

5.2.4 Mass Transport Through Cylinders

The 2-D sphere-in-cell models are completely equivalent to 2-D cylinder-in-cell models. For this reason, the solutions of the previous Sect. 5.2.3 are valid for both cases.

5.2.5 Mass Transport Through Spheroids

Spheroidal geometry was more difficultly approached until the 1990s than spherical one because the flow field had to be estimated numerically, even for creeping flow conditions. This is because the governing Eq. 3.9 in terms of stream function is not separable in spheroidal coordinates whereas it is separable in Cartesian, cylindrical and spherical ones. In 1994, Dassios et al. [14] overcome this barrier by introducing the method of variable semi-separation, thus obtaining analytical expressions for the velocity components when either Happel or Kuwabara type boundary conditions were considered. (For a detailed description of the spheroid-in-cell model, see Sect. 3.2.4.)

By considering this cell approach, the governing equation for steady state mass transport in the fluid phase within the model can be written in prolate spheroidal coordinates (η, θ) as:

$$u_\eta \frac{\partial c_A}{\partial \eta} + u_\theta \frac{\partial c_A}{\partial \theta} = \frac{D}{\alpha\sqrt{\sinh^2 \eta + \sin^2 \theta}} \cdot \left(\frac{\partial^2 c_A}{\partial \eta^2} + \coth \eta \frac{\partial c_A}{\partial \eta} + \frac{\partial^2 c_A}{\partial \theta^2} + \cot \theta \frac{\partial c_A}{\partial \theta} \right) \tag{5.34}$$

where u_η and u_θ are the $\eta-$ and $\theta-$velocity components, depend on the model considered (Happel or Kuwabara type boundary conditions).

The axial symmetry of the mass transfer problem is assured by the boundary conditions (5.18) and (5.19), where the continuity of concentration for any Peclet number is described by the boundary condition at the outer boundary of the cell [10]:

$$c_A(\eta = \eta_\beta, \theta = \pi) = 1 \tag{5.35}$$

$$\left.\frac{\partial c_A}{\partial \eta}\right]_{\eta=\eta_\beta} = 0, \quad 0 \leq \theta < \pi \tag{5.36}$$

Again several approaches for describing the sorption mechanism can be adopted (see Sect. 5.2.2), as follows:

$$[(5.16)\ \text{neutral surface}] \Rightarrow \left.\frac{\partial c_A}{\partial \eta}\right]_{\eta=\eta_a} = 0, \quad 0 \leq \theta < \pi \tag{5.37}$$

$$[(5.8)\ \text{instantaneous adsorption}] \Rightarrow c_A(\eta = \eta_a, \theta) = 0, \quad 0 \leq \theta \leq \pi \tag{5.38}$$

$$[(5.9)\ \text{Langmuir isotherm}] \Rightarrow \frac{D}{\alpha\sqrt{\sinh^2 \eta_\alpha + \sin^2 \theta}} \left.\frac{\partial c_A}{\partial \eta}\right]_{\eta=\eta_a} = \frac{k}{K} c_s, \quad 0 \leq \theta \leq \pi \tag{5.39}$$

$$[(5.13)\ \text{adsorpionndash;reactionndash;desorption}] \Rightarrow \frac{D}{\alpha\sqrt{\sinh^2 \eta_\alpha + \sin^2 \theta}} \left.\frac{\partial c_A}{\partial r}\right]_{r=\alpha} = R_n, \quad 0 \leq \theta \leq \pi \tag{5.40}$$

As shown above, the consideration of high *Pe* numbers corresponds to the analytical solution to the mass transport problem [8, 10]. The assumption of a very thin boundary layer leads to a transformed formation of Eq. 5.34 as follows:

$$u_\eta \frac{\partial c_A}{\partial r} + u_\theta \frac{\partial c_A}{\partial \theta} = \frac{D}{\alpha\sqrt{\sinh^2 \eta_\alpha + \sin^2 \theta}} \frac{\partial^2 c_A}{\partial \eta^2} \tag{5.41}$$

which can be solved analytically in a manner similar to that of Levich [17], providing concentration profiles in the fluid phase of the form given by (5.28) while the transformed variable z is given as:

$$z = \left[\frac{\alpha \mathrm{K} Pe}{4\Lambda \sinh^2 \eta_a}\right]^{\frac{1}{3}} \eta \frac{\sin \theta}{\left[\frac{\sin 2\theta}{4} - \frac{\theta}{2}\right]^{\frac{1}{3}}} \tag{5.42}$$

where K and Λ are given by Coutelieris et al. [8]. Whenever Peclet attains low or moderate values, a numerical approach must be involved, as discussed previously.

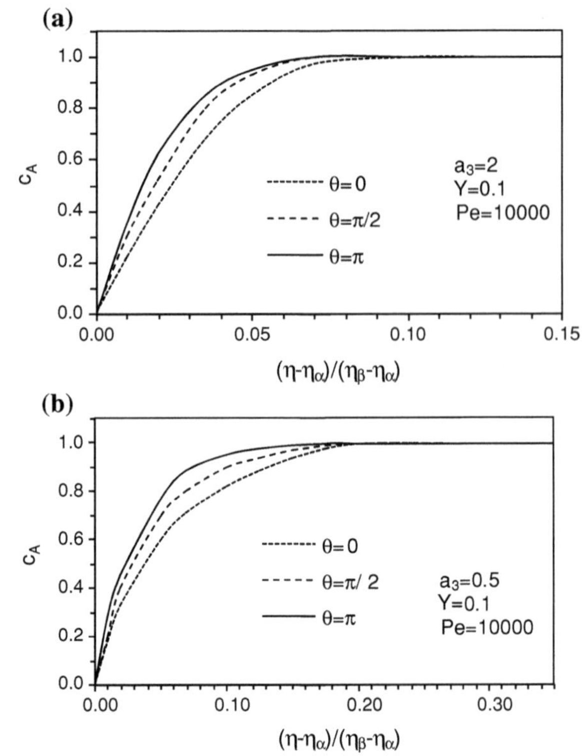

Fig. 5.9 Concentration profiles in the vicinity of adsorbing prolate (**a**) and oblate (**b**) spheroids-in-cell at three different angular positions for instantaneous adsorption and high Pe

By employing the transformation for the oblate spheroidal coordinates, given by (3.38) and (3.39), the concentration profile is again described by (5.28), where:

$$z = \left[\frac{\bar{\alpha}\,\bar{\mathrm{K}}\,Pe}{4\bar{\Lambda}\left(\sinh^2\bar{\eta}_a+1\right)}\right]^{\frac{1}{3}}\bar{\eta}\,\frac{\sin\theta}{\left[\dfrac{\sin 2\theta}{4}-\dfrac{\theta}{2}\right]^{\frac{1}{3}}} \tag{5.43}$$

and $\bar{\mathbf{K}}$, $\bar{\boldsymbol{\Lambda}}$ are given by Coutelieris et al. [8].

Again the assumption of neutral solid grains clearly results in a uniform concentration profile that lacks scientific interest. For instantaneous adsorption and using Kuwabara's approach for the velocity field, the concentration profiles for a high Peclet value (Pe = 10,000) are presented in Fig. 5.9.

The assumption of instantaneous adsorption for *low* Peclet values corresponds to the concentration profiles in Fig. 5.10

For Langmuir type adsorption with the same geometry and flow conditions, the concentration profiles are shown in Fig. 5.11.

Figure 5.12 depicts the concentration profiles for the realistic sorption mechanism with the same geometry and flow conditions as above.

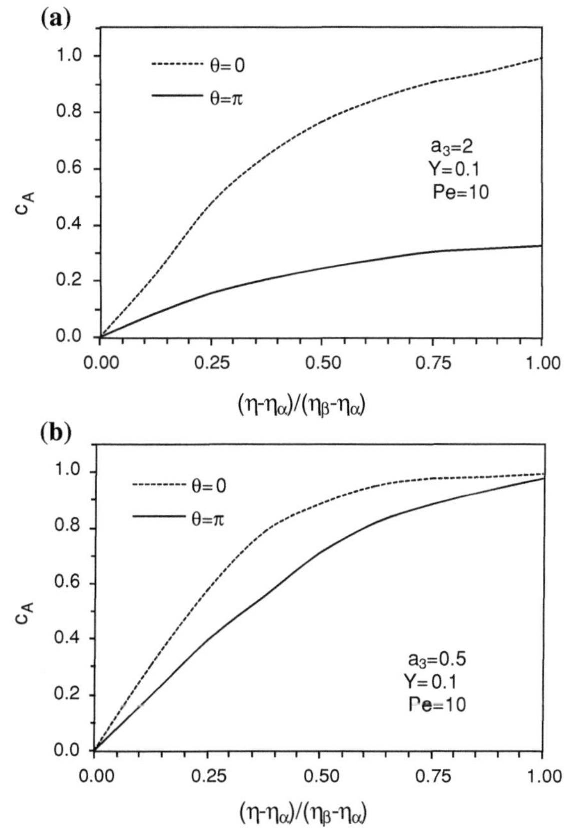

Fig. 5.10 Concentration profiles in the vicinity of adsorbing prolate (**a**) and oblate (**b**) spheroids-in-cell at three different angular positions for instantaneous adsorption and low Pe

Clearly, the results are qualitatively similar to those for spheres with the same transport and adsorption conditions.

For all the sorption mechanisms considered, and for the case of prolate spheroidal coordinates, the overall Sherwood number, Sh_o, is calculated as follows [9]:

$$Sh_o = \frac{1}{\left(1 + \frac{a_3}{a}\sin^{-1}\frac{a}{a_3}\right)} \int_{\pi}^{0} \left(\frac{\partial c_A}{\partial \eta}\right)_{\eta=\eta_a} \sin\theta \, d\theta \tag{5.44}$$

and, for oblate spheroidal coordinates [9]:

$$Sh_o = \frac{1}{\left(1 + \frac{\bar{a}_3^2}{2\bar{a}}\ln\frac{1+\bar{a}}{1-\bar{a}}\right)} \int_{\pi}^{0} \left(\frac{\partial c_A}{\partial \bar{\eta}}\right)_{\bar{\eta}=\bar{\eta}_a} \sin\theta \, d\theta \tag{5.45}$$

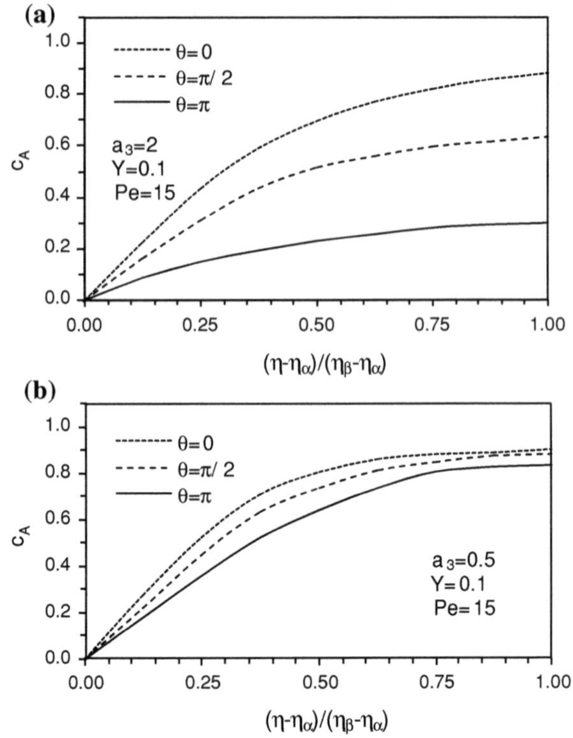

Fig. 5.11 Spatial concentration profiles at three different angular positions of the prolate (**a**) and oblate (**b**) spheroid-in-cell for Langmuir type adsorption $\left(\frac{k}{K} = 1\right)$

Finally, the adsorption efficiency of prolate cells can be written as:

$$\lambda_0 = \frac{2}{Pe\, b_1^2} \int_{\pi}^{0} \left(\frac{\partial c_A}{\partial \eta} \right)_{\eta = \eta_a} \sin \theta \, d\theta \tag{5.46}$$

and

$$\bar{\lambda}_0 = \frac{2}{Pe\, \bar{b}_1^2} \int_{\pi}^{0} \left(\frac{\partial c_A}{\partial \bar{\eta}} \right)_{\bar{\eta} = \bar{\eta}_a} \sin \theta \, d\theta \tag{5.47}$$

The effect of the Peclet number on the Sherwood number for instantaneous adsorption is presented in Fig. 5.13.

As shown above, the boundary condition on the outer boundary is crucial for the estimation of macroscopic quantities such as the overall Sherwood number and adsorption efficiency. The dashed lines in Fig. 5.13 correspond to a "Levich-equivalent" approach, where a constant uniform concentration profile is considered on the surface of the outer spheroid. This condition is compatible with the assumption of a thin boundary layer (i.e. high Peclet regime), unless it breaks down when diffusion becomes significant and the boundary layer thickens. Indeed,

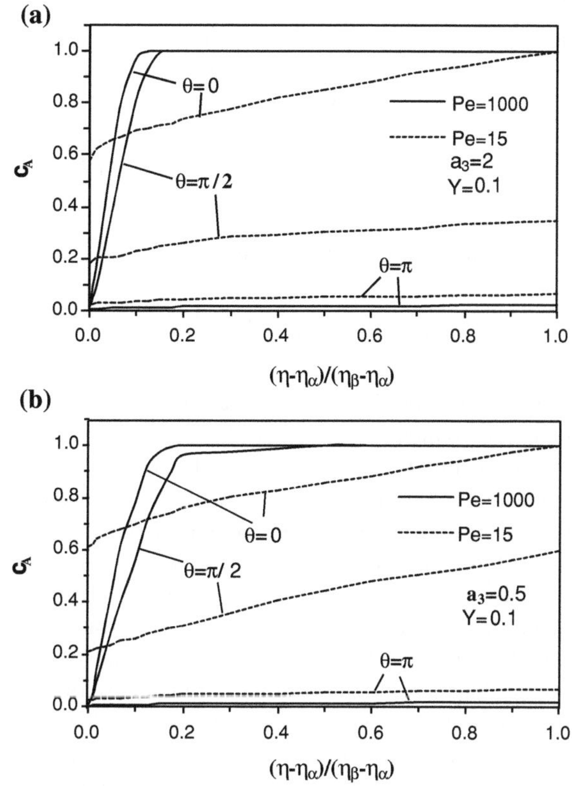

Fig. 5.12 Concentration profiles for prolate (**a**) and oblate (**b**) spheroidal shapes for high and low Peclet numbers at three different angular positions

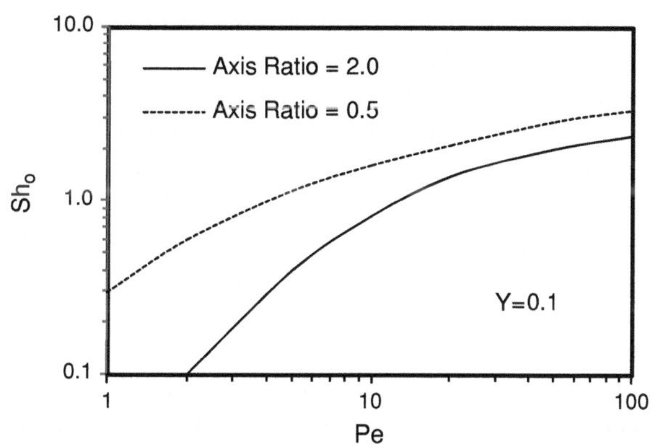

Fig. 5.13 Dependence of the overall Sherwood number on the Peclet for prolate and oblate spheroids-in-cell (instantaneous adsorption)

for $Pe \geq 20$ the Levich type approach predicts the correct Sh_o value with sufficient accuracy while, for lower Pe values, it yields Sh_o values weakly dependent on Pe, and converges to a limiting value as $Pe \to 0$. This limiting value is a function of porosity and the axis ratio, given by [9]:

$$Sh_0^* = 2a\left(1 + \frac{a_3^2}{a}\sin^{-1}\frac{a}{a_3}\right)^{-1}\left(\ln\frac{a+b_3}{b_1(a+a_3)}\right)^{-1} \tag{5.50}$$

for prolate spheroids-in-cell, and by:

$$Sh_0^* = 2\bar{a}\left(1 + \frac{\bar{a}_3^2}{2\bar{a}}\ln\frac{1+\bar{a}}{1-\bar{a}}\right)^{-1}\left(\tan^{-1}\frac{\bar{b}_3}{\bar{a}} - \tan^{-1}\frac{\bar{a}_3}{\bar{a}}\right)^{-1} \tag{5.51}$$

for oblate spheroids-in-cell. The above equations result from solving the Laplace equation, which is valid for the pure diffusive regime, i.e. as $Pe \to 0$ [5, 9]. However, conditions on the outer boundary (5.35) and (5.36) are obviously more consistent, resulting in more realistic Sh_o. For the realistic adsorption mechanism, typical analogous results are shown in Fig. 5.14.

In Fig. 5.14, a monotonic decrease of Sh_o, with increasing aspect ratio, is observed for all cases because prolate spheroids present higher diffusional resistance than oblate ones [6, 8]. Oblate spheroids present larger impact surface than prolate ones, and thus their capacity for adsorption is higher. This advantage of oblate shapes becomes very weak in the case of small Pe values because almost all the parts of the adsorbing solid surface become active as diffusion becomes dominant. For small Peclet values, unlike large ones, the overall Sherwood number depends appreciably on the order of the heterogeneous reaction because the concentration gradients become less significant. In this case, a decrease of Sh_o is observed when the reaction is of second-order (rather than first-order) as a higher reaction order causes lower concentration gradients and larger amounts of adsorbed mass. The overall Sherwood number can be 15–50% higher, depending on Pe, the shape, porosity, and order of the reaction, compared to the values obtained assuming instantaneous adsorption for either high or low Peclet values [8, 9]. This occurs because the concentration on the solid surface attains nonzero values and, thus, the difference $c_A(\eta_\beta, \theta) - c_A(\eta_a, \theta)$ becomes almost nil for some θ-values. The decrease of this driving force is significantly larger than the decrease of the concentration gradients observed when a realistic adsorption process is assumed rather than an instantaneous one.

Figure 5.15 presents adsorption efficiency results for prolate and oblate spheroids-in-cell, as functions of the Peclet number. The Levich approach on the outer boundary overestimates the adsorption efficiency for low Peclet values, leading to values larger than unity as $Pe \to 0$ for both geometries. On the contrary, the modification (5.35) and (5.36) predict λ_o values, which remain lower than unity even in the very low Pe range. What is observed in general is that convention is a favorable parameter allowing the escape of diluted material from the adsorbing solid surface.

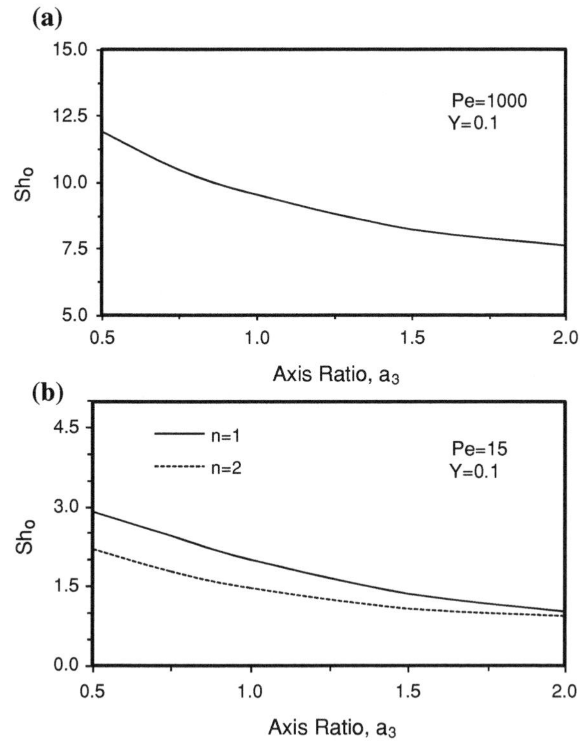

Fig. 5.14 Dependence of the overall Sherwood number on the axis ratio of the spheroid for a high (**a**) and a low (**b**) Peclet number (realistic adsorption)

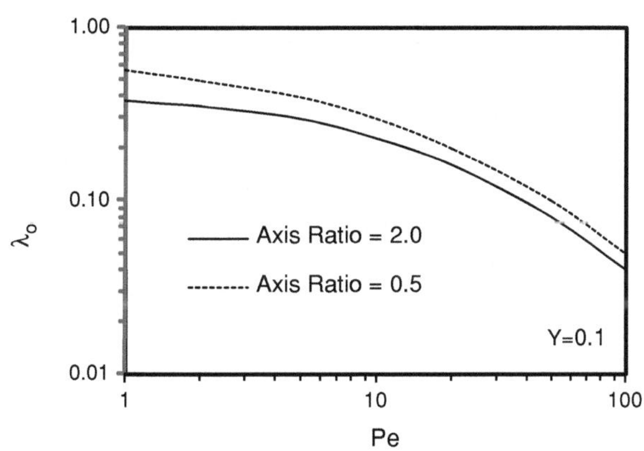

Fig. 5.15 Dependence of the adsorption efficiency on the Peclet number for prolate and oblate spheroids-in-cell (instantaneous adsorption)

The dependence of the overall adsorption efficiency on the Peclet number for both prolate (a) and oblate (b) geometry is given in Fig. 5.16, where the heterogeneous reaction is assumed to be either of first ($n = 1$) or second ($n = 2$) order. A significant decrease of the adsorption efficiency occurs as Pe increases. Values of overall adsorption efficiency obtained using the instantaneous adsorption model are 10–35% higher than those obtained here, because of the higher concentration gradients that instantaneous adsorption generates.

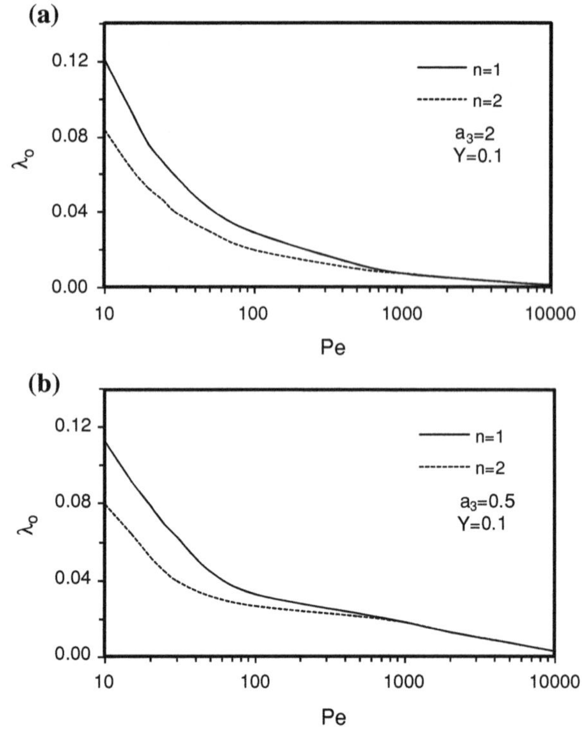

Fig. 5.16 Influence of the Peclet number on the adsorption efficiency for prolate (**a**) and oblate (**b**) spheroids-in-cell (realistic adsorption mechanism)

5.2.6 Single Phase Mass Transport in Other-Type Unit Cells

Further to the models described above, several unit cells have also been proposed in recent decades. These assemblies are all characterized by: (a) complex geometry which enforces numerical solutions of flow and transport problems, (b) the iteration of their structure to occupy the volume of interest (periodic boundary conditions), and (c) a high degree of freedom in the geometrical shapes considered to represent the real porous material. Increased computational power nowadays allows for the consideration of huge complex structures, therefore current research trends digress from the unit cell formulation, as analytical solutions no longer essential.

5.3 Single Phase Flow in Granular Structures

As described in Sect. 3.3, transport processes in granular media are of high scientific and technological interest, and mathematical modeling plays an important role in the in-depth study and understanding of these processes. In the majority of models on transport in granular media, the particles of the assemblage are assumed to adsorb mass instantaneously, which is obviously a fairly rare physicochemical phenomenon that pertains to a very limited range of applications. Only recently have several detailed models of the sorption mechanism been presented for porous media processes [10, 13].

The aim of this Section is to present an adequate simulation for the adsorption heterogeneous reaction–desorption mechanism, which accurately describes the sorption upon a solid surface of a solute diluted in a flowing fluid [2, 20]. More specifically, it can be considered that the solute diluted in the bulk phase is initially adsorbed by the solid surface where a heterogeneous reaction takes place and its products, assumed to be inactive and of very low concentrations, are again desorbed in the bulk phase. The adsorption is assumed to occur due to vacant sites that are normally distributed over the surface area, while the whole process is determined by an overall rate according to the flow regime and thermodynamics [25].

As described in Sect. 3.3, the procedure of random deposition under specific restrictions must be followed in order to represent the stochastically constructed three-dimensional assemblages of spheres which are considered a typical demonstration of a granular medium (see Fig. 3.7). The flow-field within such an assemblage is typically obtained by the numerical solution of the Stokes equations coupled with the continuity equation, as described in Sect. 3.3.

Assuming no reactions occur in the bulk phase, the time-dependent mass transport of a passive solute (namely, component A) in the stochastically constructed medium is described by the convection–diffusion equation:

$$\frac{\partial c_A}{\partial t} + \nabla \cdot (\mathbf{u} c_A) = D \nabla^2 c_A \tag{5.52}$$

To investigate the importance of the adsorption mechanism in such complex structures, all the previously discussed different adsorption mechanisms are adopted here. Neutral solid surfaces again lack scientific and engineering interest. The other sorption mechanisms considered are instantaneous adsorption (5.8), adsorption under the Langmuir isotherm (5.9)–(5.11), and the realistic sorption mechanism (5.13)–(5.15). Typical conditions are also considered at the other boundaries.

The validity of the solution scheme for the mass transport problem can be checked in the limiting case of a periodic array of spheres with radius R, which is fed at $x = 0$ by a concentration pulse. The concentration profile with time at the outlet $\left(x = L = \frac{51}{8} R\right)$ is then expressed as [23]:

$$c(t) = \frac{1}{\sqrt{4\pi D^* t}} \exp\left[\frac{-(L - \langle u \rangle t)^2}{4 D^* t}\right] \tag{5.53}$$

where $\langle u \rangle$ is the average fluid velocity, and D^* the dispersion coefficient. Therefore, the dimensionless time (defined by using $\frac{L^2}{D^*}$ as characteristic time) at which the maximum outlet concentration is recorded can be calculated as:

$$t_{\max} = \frac{\left(\sqrt{1 + Pe^2} - 1\right)}{Pe^2} \tag{5.54}$$

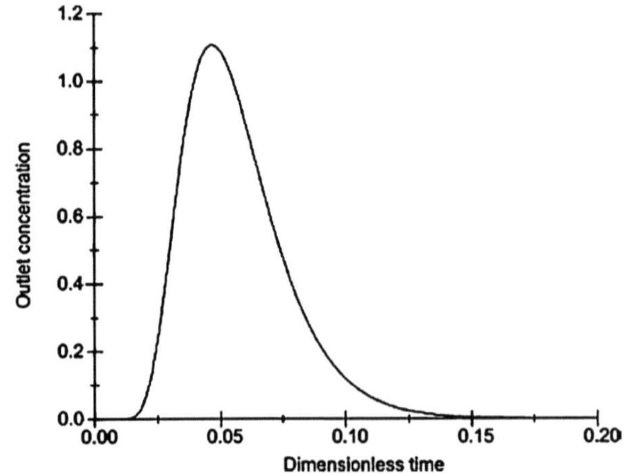

Fig. 5.17 Outlet concentration profile for Pe = 20 in a periodic array of spheres

where $Pe = \dfrac{\langle u \rangle L}{D}$ is the Peclet number.

The dispersion coefficient can be determined based on the approach of Salles et al. [24] and in these simulations is found to be $D^* = 1.34D$, thus giving $t_{\max} = 0.047$ for $Pe = 20$, when the numerically calculated value is 0.048 (see Fig. 5.17).

Note that the dimensionless time step dt must be small enough ($<10^{-5}$) to attain acceptable accuracy in the calculations.

The spatial distribution of the concentration, obtained numerically for $Pe = 20$ and $\varepsilon = 0.7243$ along the flow direction, is presented in Fig. 5.18 where a randomly selected two-dimensional cut of the stochastically constructed medium is visualized for both instantaneous and Langmuir-type adsorption. The concentration is higher on the left side of the images (inlet) and gradually decreases across the medium in a manner significantly steeper for the instantaneous than the Langmuir-type adsorption. The adsorption efficiency for several porosities ($\varepsilon = 0.9883$, $\varepsilon = 0.8136$ and $\varepsilon = 0.7243$) and Peclet numbers was calculated by the full-numerical scheme in the sphere assembly for both instantaneous and Langmuir-type adsorption and presented in Fig. 5.19. The effects of porosity and Peclet on the adsorption efficiency are the same as those above (higher efficiencies correspond to lower porosity, and increasing Peclet numbers lead to lower λ_O values). Instantaneous adsorption leads to higher adsorption efficiency values than those calculated under other adsorption mechanisms with the same Peclet number and porosity.

The effect of the k/K ratio used in the boundary condition for Langmuir-type adsorption, on the computed adsorption efficiency is presented in Fig. 5.20 for $Pe = 20$ and $\varepsilon = 0.8136$. As this ratio increases, a significant increase of λ_O is observed because higher k/K corresponds to lower ratios of the covered surface, Θ_{eq}. This, in turn, corresponds to lower surface concentrations and higher concentration gradients, i.e. to higher λ_O. On the other hand, an increased k/K ratio can be viewed as a higher adsorption rate for a given K value and therefore, less mass of substance A can escape from the solid surfaces resulting in higher λ_O values.

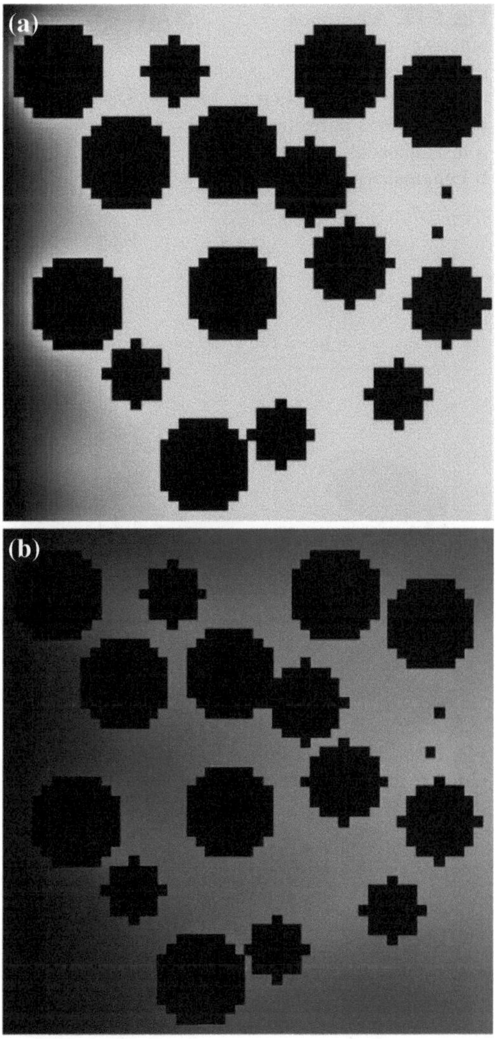

Fig. 5.18 Spatial distribution of the concentration of substance A within a two-dimensional cut of the three-dimensional sphere assemblage for **a** instantaneous and **b** Langmuir-type adsorption. (The flow inlet is on the left side. *Darker areas* correspond to higher concentrations.)

It is interesting to consider the relative agreement between the results produced by considering the sphere-in-cell approximation and those obtained from the numerical solution of the three-dimensional sphere assemblages. Figure 5.21a compares the respective adsorption efficiencies and instantaneous adsorption for porosity approaching unity ($\varepsilon = 0.9883$). The agreement is perfect as the semi-analytical sphere-in-cell model can adequately predict reality for such high porosity values. When lower porosities are considered, the fundamental assumptions of the sphere-in-cell approximation are less satisfactory. Indeed, in the low *Pe* regime characterized by the gradual dominance of the diffusive over the convective terms, the semi-analytical approach of the sphere-in-cell model cannot adequately describe the mass transport process, as the diffusion layer is very thick

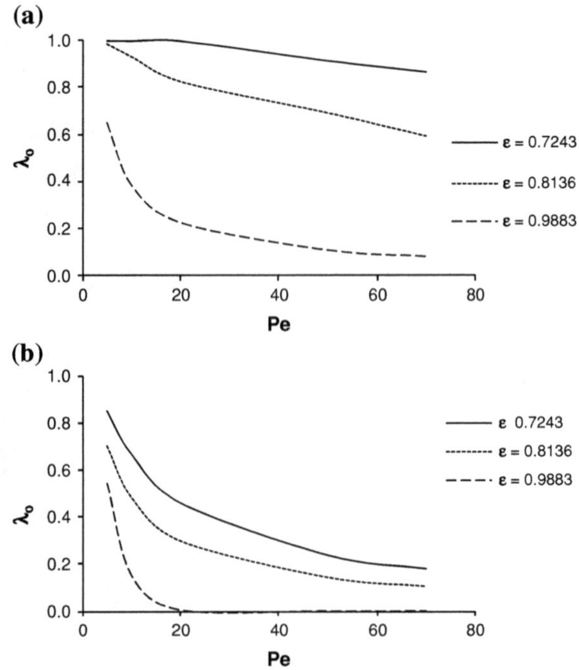

Fig. 5.19 Adsorption efficiency of the three-dimensional sphere assemblage as a function of Pe for various porosities in **a** instantaneous and **b** Langmuir-type adsorption

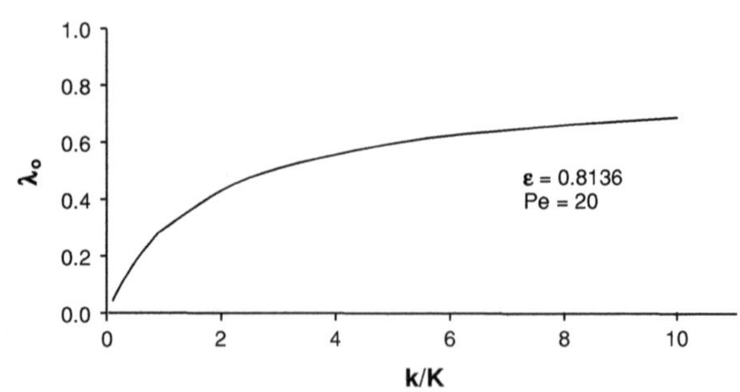

Fig. 5.20 The effect of k/K ratio on adsorption efficiency

(larger than the cell itself) and tends to infinity as $Pe \to 0$. On the other hand, as Pe increases, flow becomes more convective and the analytical flow-field of the sphere-in-cell is no longer a sufficient approximation of the actual flow-field in real granular media, thus leading to model discrepancies. However, for $\varepsilon = 0.8136$ and $\varepsilon = 0.7243$ the agreement between the two approaches ranges from very good to satisfactory depending on the Pe range considered. This result is expected since both the sphere-in-cell and sphere assemblage geometries used in the present study were constructed to be characterized by the same porosity and internal (adsorbing) surface area. This was accomplished by selecting the correct radius and sphere population number in the assemblage. Of course, it is possible to construct several sphere assemblages of the same porosity but with varying sphere radius and

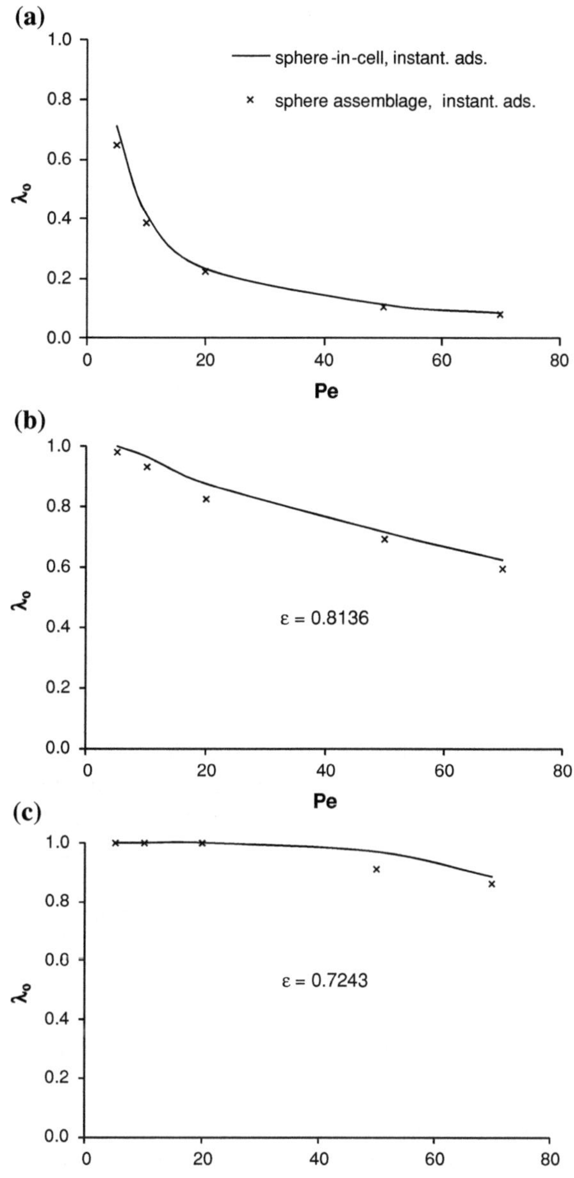

Fig. 5.21 Comparison between adsorption efficiency in sphere-in-cell and sphere assemblies for different porosities

population number and therefore with different internal surfaces. Figure 5.22 shows how these different arrangements of spheres affect the calculated adsorption efficiency for instantaneous and Langmuir adsorption. The horizontal axes represent the internal surface ratio which changes when different media are considered. The case of adsorption surface ratio equal to unity corresponds to the results discussed so far. Evidently under instantaneous adsorption conditions λ_0 is not

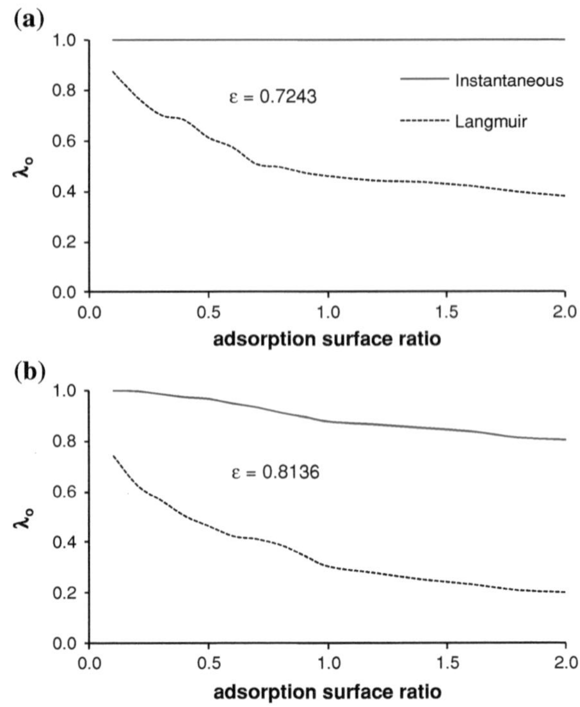

Fig. 5.22 The effect of using sphere assemblages with varying internal surface area on adsorption efficiency and for two porosity values: 0.7243 (**a**) and 0.8136 (**b**)

influenced significantly by changes in the medium. This is again due to the high concentration gradients prevailing that mask the effect of internal surface variation. However, in the more realistic Langmuir-type of adsorption a very pronounced influence on λ_0 is observed implying that caution should be exercised when using the sphere-in-cell model to determine adsorption efficiency in granular media. This strongly indicates that matching porosity alone is not sufficient for a reliable result.

5.4 Macroscopic Quantities for Single Phase Transport

There are many industrial and technological applications of mass transport within porous media in a variety of scientific fields, such as environment, energy, biology, etc. [15, 18, 27]. Mathematical modeling of transport processes in porous med,ia is a powerful tool, especially whenever experimental observations are difficult, time consuming and expensive. Due to the coupling between the physicochemical mechanisms and the local geometry of the porous medium mathematical descriptions of mass transport in realistic porous media are highly complex. Modeling becomes more difficult when moving from the pore level to the field level, because different length scales result in complicated descriptions of the problem's physics and therefore increased computational power is usually required.

From the late 1950s, special effort has been given to mathematically describe and solve flow and mass transport problems in porous media. Initially quite simplified geometries were considered, where analytical solutions can be obtained for the flow field and the mass diffusion and/or adsorption process (see previous chapters on cell models). Numerical solutions in realistic reconstructions of porous media for the Stokes equations and related transport problems have been obtained for several specific applications during the last decades [1, 3, 16, 19, 26]. In the majority of these works, the particles were assumed to adsorb mass instantaneously however, this is a rare physicochemical phenomenon that pertains to a very limited range of applications. Recently, further detailed models of the sorption mechanism have been presented for porous media processes.

This Section aims to simulate an adsorption—heterogeneous reaction—desorption mechanism for a complex granular porous geometry that can accurately describe the sorption upon a solid surface of a solute diluted in the flowing fluid [2, 20]. Details about this sorption mechanism can be found in Sect. 5.2.

The major issue of typical macroscopic modeling for such, or simpler, cases can be identified at the a priori definition of the macroscopic quantities necessary to solve these equations, although they are normally derived from the solution of these equations. So far, mainly empirical or semi-empirical correlations for these parameters have been proposed based on experimental measurements of specific systems [5, 22]. The generalized treatment of such a problem corresponds to theoretical estimations of these quantities where the volume averaging concept is a frequently employed tool for large-scale modeling of processes taking place in porous media, thus eliminating the influence of porous geometry on the transport results [28, 30, 31]. Starting with transport equations at the micro-scale (pore) level, the spatial averaging theorem is applied with the correct assumptions, leading to the estimation of macroscopic quantities such as mass transfer coefficient and dispersion tensor [21]. To further simplify the modeling and eliminate the simulation effort, the majority of the above-mentioned models have been applied to simplistic domains, such as unit cells, since the focus was on the interfacial mass exchange rather than the representation of the medium in a realistic manner.

5.4.1 *Stochastically Constructed 3-D Sphere Assemblage*

Representation of the biphasic domains under consideration is achieved by the random deposition of spheres of a given radius in a box of specified length. (For specific details on the efficient algorithm (ballistic deposition) used for the domain's stochastic construction, see Sect. 3.3).

5.4.2 *The Flow Field (Single Phase Flow)*

The velocity field is computed numerically by solving the Stokes equations as described previously (see Sect. 3.3).

5.4.3 Mathematical Formulation

The area of interest is a porous domain consisting of a flowing aqueous phase (*L*-phase) and a solid phase (*S*-phase). A tracer is advected by the flowing liquid phase, being sorpted in the liquid–solid interface. The governing processes are diffusion and advection in the liquid phase, and the liquid–solid interface is characterized by the realistic adsorption/reaction/desorption mechanism of the tracer (see below).

Assuming that the bulk phase is chemically neutral, the pore-level transport of the tracer in the β-phase is described by the convection–diffusion equation:

$$\frac{\partial C_L}{\partial t} + \nabla \cdot (\mathbf{v} C_L) = D_L \nabla^2 C_L \tag{5.55}$$

where C_L is concentration, t is time, $\mathbf{v}$ is the fluid velocity, and D_L is the diffusivity in the liquid phase.

To ensure the continuity of the mass fluxes on the solid–liquid interfaces, the following boundary condition (similar to Eq. 5.13) is applied:

$$\mathbf{n}_{LS} \cdot \nabla C_L = R_n, \text{ at the solid–liquid interface } A_{LS} \tag{5.56}$$

where the overall sorption rate R_n is dependent on the type of sorption process considered. In accordance to Sect. 5.2, the rate R_n is given as:

$$R_n = k_s C_{LS}^n \tag{5.57}$$

where the concentration of the tracer upon the solid surface, C_{LS}, is described by the relation:

$$k_s C_{LS}^n + [k_d + k_a C_L N] C_{LS} - k_a C_L \xi_m = 0 \tag{5.58}$$

5.4.4 The Volume-Averaging Procedure

Following the volume-averaging procedure [22], local concentrations and velocities are next decomposed into interstitial averages and fluctuations

$$C_L = \langle C_L \rangle^L + C_L' \tag{5.59}$$

$$\mathbf{v} = \langle \mathbf{v} \rangle^L + \mathbf{v}' \tag{5.60}$$

which are subsequently substituted in the governing differential equations. Invoking separation of scales to discard small terms, linearizing and following Quintard and Whitaker [22], assuming an isotropic medium of uniform porosity and constant volume fractions, the following representation is obtained:

$$C_L' = \mathbf{b} \cdot \nabla \langle C_L \rangle^L - s_L \langle C_L \rangle^L \tag{5.61}$$

where $\mathbf{b}$ and s_L are closure variables satisfying specific boundary value problems.

It can be readily shown that variable s of the β-phase satisfies the boundary value problem:

$$\mathbf{v} \cdot \nabla s_L = D\nabla^2 s_L - \varepsilon^{-1}\alpha \tag{5.62}$$

with

$$\mathbf{n}_{LS} \cdot \nabla s_L = 0, \text{ at the solid–liquid interface } A_{LS} \tag{5.63}$$

and

$$\langle s_L \rangle = 0 \tag{5.64}$$

The mass-transfer coefficient, α, on Eq. 5.62, is given by:

$$\alpha = \frac{D}{V} \int\limits_{A_{LS}} \mathbf{n}_{LS} \cdot \nabla s_L dA \tag{5.65}$$

The above can be simplified by introducing the transformation $s_L = 1 + \alpha\psi_{L\sigma}$ and in dimensionless form:

$$s_L = 1 + \alpha^* \zeta_L \tag{5.66}$$

where the dimensionless mass-transfer coefficient is defined as:

$$\alpha^* = \frac{a l_L^2}{D} \tag{5.67}$$

and l_L denotes the characteristic length of the liquid phase. In dimensionless notation, the boundary value problems then read as follows:

$$Pe\, \mathbf{u} \cdot \nabla \zeta_L = \nabla^2 \zeta_L - \varepsilon_L^{-1} \text{ in the liquid phase} \tag{5.68}$$

$$\mathbf{n}_{LS} \cdot \nabla \zeta_L = r_n \text{ at the solid–liquid interface } A_{LS} \tag{5.69}$$

where $Pe = \frac{\langle \mathbf{u} \rangle^L l_L}{D}$ is the Peclet number defined in the liquid phase, $\mathbf{u}$ is the dimensionless velocity vector, ζ_L is the scalar variable used for the decomposition, and r_n is the dimensionless sorption rate.

Following the above, the dimensionless mass transfer coefficient simply becomes:

$$\alpha^* = -\frac{\varepsilon}{\langle \zeta_L \rangle} \tag{5.70}$$

Brackets denote averages over the total volume V or the volume of the aqueous phase, V_L, where the superficial volume average is defined as:

$$\langle y_L \rangle = \frac{1}{V} \int\limits_{V_L} y_L dV \tag{5.71}$$

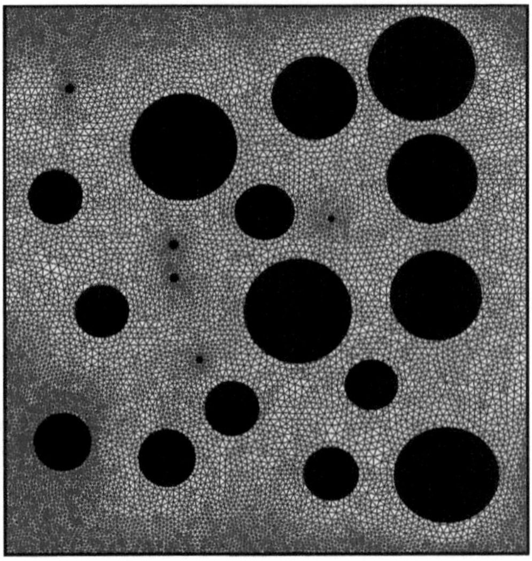

Fig. 5.23 A selected two-dimensional cut of the simulated geometry discretized by an unstructured grid

and the interstitial volume average as:

$$\langle y_L \rangle^L = \frac{1}{V_L} \int_{V_L} y_L dV \tag{5.72}$$

5.4.5 Simulations

To adequately simulate the above-described problem, an algorithmic procedure has been developed as follows:

- Solve the flow problem at the pore level and calculate interstitial and superficial velocity fields
- Formulate the mass transport problem at the pore level
- Decompose the local velocity and concentration in terms of an interstitial average and a fluctuation
- Describe the concentration fluctuations in terms of linear combinations of interstitial averaged concentration and its gradient
- Solve the closure problem
- Integrate the resulting quantities to calculate macroscopic coefficients.

The numerical scheme used for all the simulations is described in Sect. 3.3 and a typical grid representation is depicted in Fig. 5.23 (for the same two-dimensional cut as the previous Fig. 5.18).

Table 5.1 Independence on random deposition ($\varepsilon = 0.43$).

Different random assemblages	*Pe*	*n*	a^*
Case 1	1	1	20.93
Case 2	1	1	20.83
Case 3	1	1	21.01
Case 4	1	1	20.77
Case 5	1	1	20.96
Case 1	100	1	46.55
Case 2	100	1	45.87
Case 3	100	1	46.11
Case 4	100	1	47.02
Case 5	100	1	46.39
Case 1	1	2	19.36
Case 2	1	2	19.91
Case 3	1	2	19.07
Case 4	1	2	18.66
Case 5	1	2	19.12
Case 1	100	2	45.88
Case 2	100	2	45.01
Case 3	100	2	46.99
Case 4	100	2	45.85
Case 5	100	2	46.02

5.4.6 Results and Discussion

The results were initially validated against the randomness of the structure, as previously discussed. More precisely, the mass transfer coefficient a^* was calculated for several different depositions of spheres while the porosity value was kept constant (=0.43). It was found that the random deposition does not significantly affect mass transfer to the solid phase, as shown in Table 5.1, where each "case" corresponds to a different deposition of a variable number of spheres of different radii, and porosity is kept constant in all cases. (Five different cases were chosen for the same porosity to verify whether the results were independent of the medium construction). Solution dependence on the grid was also examined in terms of the mass transfer coefficient. It was found that the discretization used is more than sufficient for adequate calculations. It should be noted that the parametric analysis of the grid influence is limited to only reaction order, as this parameter introduces non-linearity, thus the solution it is depended on this reaction order.

The relative agreement between the results produced by considering the approximation presented in Sect. 5.3 and those obtained using the above technique is presented in Fig. 5.24. By assuming a typical value of ξ_m (1 active site per Å^2), the values of the sorption constants were (before the non-dimensionalization of the problems) $k_a = 1 \times 10^{-30}\ m^3 sec^{-1}$, $k_d = 8 \times 10^{-3}\ sec^{-1}$ and $k_s = 8 \times 10^{-3} \times 100^{(1-n)}$ $(kg\ m^{-2})^{1-n} sec^{-1}$. These values can be considered as typical [2] and are used in the

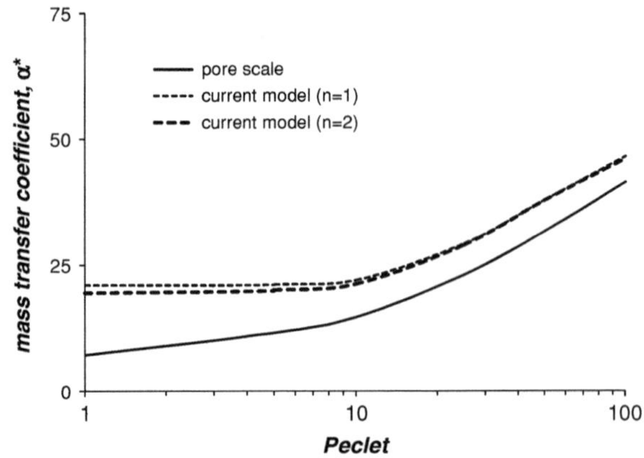

Fig. 5.24 Comparison between the results of the current model and those found in the literature

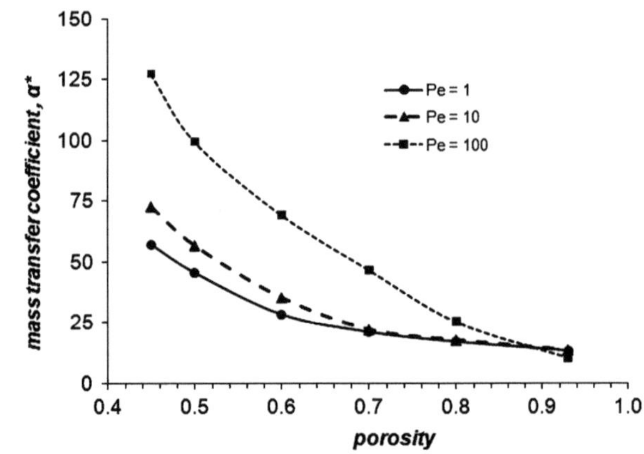

Fig. 5.25 The influence of porosity on mass transport

simulations presented here unless otherwise stated. Figure 5.24 compares the respective mass transfer coefficient for the standard porosity $\varepsilon = 0.43$ while the realistic adsorption/reaction/desorption mechanism includes the heterogeneous reaction of first ($n = 1$) and second ($n = 2$) order. Regarding the pore-level simulations, a discrepancy from the results of the current model is always observed, thus indicating the underestimation of macroscopic mass transport quantities when calculated using pore-level approaches in small-scale domains [7].

In terms of physical interpretation, Fig. 5.24 depicts the effect of convection on mass transport. It can be seen that the stronger the convection, the more efficient the tracer transport from the fluid to the solid phase, at least for low and intermediate porosity values, which correspond to relatively high amounts of active solid absorbers in the medium and absorb the tracer. Finally, the order of the reaction does not seem to significantly affect the results, or the agreement between the two approaches. Only the first order reaction is considered hereafter.

The influence of the medium's volume porosity on mass transport is presented in Fig. 5.25. It is clear that porosity is an unfavorable parameter for adsorption,

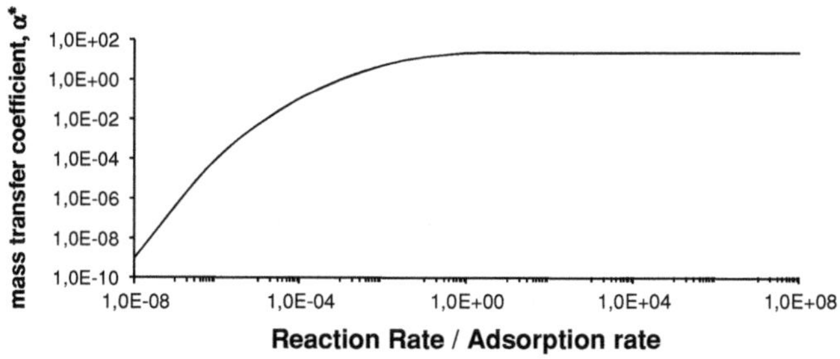

Fig. 5.26 Dependence of mass transport on the ratio of the reaction to the adsorption rates

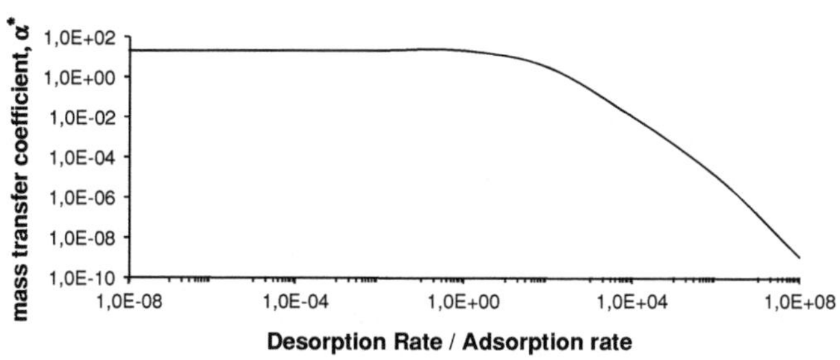

Fig. 5.27 Dependence of mass transport on the ratio of desorption to adsorption rates

since void space increases as porosity increases (although the active solid surface area does not necessarily decrease), thus corresponding to high possibility for the tracer to escape from the porous material through the void space. For high porosity values, as Peclet values increase, the transport process becomes more and more convective, thus mass transport becomes less effective, i.e. high amounts of the tracer can escape from the medium. On the other hand, low porosity values correspond to large amounts of solid phase in the medium, thus convection favors mass transport since the tracer is forced to approach the absorbing surfaces. These two competitive phenomena are shown by the cross of the curves in Fig. 5.25, which correspond to a porosity value where both mechanisms are of equal strength. This value obviously depends on the specific flow and transport characteristics.

Figures 5.26 and 5.27 depict the relative influence of the sorption mechanism, i.e. the reaction, adsorption and desorption rates, on mass transport. More precisely, Fig. 5.26 presents the mass transfer coefficient as a function of the ratio of tracer destruction rate due to the reaction, divided by the tracer destruction rate due to adsorption. In any case, the values of the rate constants not involved in these ratios were kept standard. It is observed that the decrement of the reaction rate (for constant adsorption rate) corresponds to a consequent decrement of mass transport because the tracer has been adsorbed but not destroyed at the same rate and, therefore remains on the surface filling the vacant sites, i.e. setting barriers in the tracer sorption process. The increased reaction rate forces adsorption to tend asymptotically to a constant value which depends on the geometrical characteristics of the medium. Figure 5.27 shows the influence of the ratio desorption/adsorption rate on mass transport for a heterogeneous reaction of the first order. It is observed that increasing the desorption rate beyond a critical value corresponds to a decrease in mass transport. It is important to note that the value attained by the mass transfer coefficient before the critical desorption rate is the same as the asymptotic value of Figure 5.26, thus further underlying its independence on the reaction characteristics.

References

1. Ahmadi, A., Aigueperse, A., Quintard, M.: Calculation of the effective properties describing active dispersion in porous media: from simple to complex porous media. Adv. Water Resour. **24**, 423–438 (2001)
2. Atkins, P., de Paula, J.: Physical Chemistry, 7th edn. Oxford University Press, Oxford (2002)
3. Bekri, S., Thovert, J.F., Adler, P.M.: Dissolution and deposition in fractures. Eng. Geo. **48**, 283–308 (1997)
4. Bird, R.B., Stewart, W.E., Lightfoot, E.N.: Transport Phenomena. Wiley, New York (1960)
5. Burganos, V.N., Coutelieris, F.A., Payatakes, A.C.: Sherwood number for mass transfer to a swarm of adsorbing spheroidal particles at any Peclet number. AIChE J. **43**, 844–848 (1997)
6. Coutelieris, F.A.: The influence of axial orientation of spheroidal particles on the adsorption rate in a granular porous medium. Stud. Surf. Sci. Catal. **144**, 745–751 (2002)
7. Coutelieris, F.A.: Modeling of flow and mass transport in granular porous media. Cent. Eur. J. Phys. **9**, 962–968 (2011)
8. Coutelieris, F.A., Burganos, V.N., Payatakes, A.C.: On mass transfer from a newtonian fluid to a swarm of adsorbing spheroidal particles for high Peclet numbers. J. Colloid Interface Sci. **161**, 43–52 (1993)
9. Coutelieris, F.A., Burganos, V.N., Payatakes, A.C.: Convective diffusion and adsorption in a swarm of spheroidal particles. AIChE J. **41**, 1122–1134 (1995)
10. Coutelieris, F.A., Burganos, V.N., Payatakes, A.C.: Model of adsorption–reaction–desorption in a swarm of spheroidal particles. AIChE J. **50**, 779–785 (2004)
11. Coutelieris, F.A., Kainourgiakis, M.E., Stubos, A.K.: The effect of the porosity on the adsorption rate in granular porous media. Stud. Surf. Sci. Catal. **144**, 753–760 (2002)
12. Coutelieris, F.A., Kainourgiakis, M.E., Stubos, A.K.: Low Peclet mass transport in assemblages of spherical particles for two different adsorption mechanisms. J. Colloid Interface Sci. **264**, 20–29 (2003)
13. Coutelieris, F.A., Kainourgiakis, M.E., Stubos, A.K.: Low to moderate Peclet mass transport in assemblages of spherical particles for a realistic adsorption–reaction–desorption mechanism. Powder Tech. **159**, 173–179 (2005)
14. Dassios, G., Hadjinicolaou, M., Payatakes, A.C.: Generalized eigenfunctions and complete semiseparable solutions for stokes flow in spheroidal coordinates. Quart. Appl. Math. **52**, 157–191 (1994)
15. Ellis, T.G., Eliosov, E., Schmit, C.G., Jahan, K., Park, K.Y.: Activated sludge and other aerobic suspended culture processes. Water Environ. Res. **74**, 385–410 (2002)
16. Kapellos, G.E., Alexiou, T.S., Payatakes, A.C.: A multiscale theoretical model for diffusive mass transfer in cellular biological media. Math. Biosci. **210**, 177–237 (2007)
17. Levich, V.G.: Physicochemical Hydrodynamics. Pentice-Hall, Englewood Cliffs (1962)
18. Meakin, P., Skjeltorp, A.T.: Application of experimental and numerical models to the physics of multiparticle systems. Adv. Phys. **42**, 1–127 (1993)

19. Mourzenko, V.V., Bekri, S., Thovert, J.F., Adler, P.M.: Deposition in fractures. Chem. Eng. Commun. **150**, 431–464 (1996)
20. Peters, M.H., Jalan, R.K., Gupta, D.: A dynamic simulation of particle deposition on spherical collectors. Chem. Eng. Sci. **40**, 723–731 (1985)
21. Quintard, M., Whitaker, S.: Transport in ordered and disordered porous media: volume averaged equations, closure problems and comparison with experiments. Chem. Eng. Sci. **48**, 2537–2564 (1993)
22. Quintard, M., Whitaker, S.: Convection, dispersion and interfacial transport of contaminants: Homogeneous porous media. Adv. Water Resour. **17**, 221–239 (1994)
23. Sahimi, M.: Flow phenomena in rocks: from continuum models to fractals, percolation, cellular automata, and simulated annealing. Rev. Mod. Phys. **65**, 1393–1534 (1993)
24. Salles, J., Thovert, J.F., Delannay, R., Prevors, L., Auriault, J.L., Adler, P.M.: Taylor dispersion in porous media. Determination of the dispersion tensor. Phys. Fluids A - Fluid **5**, 2348–2367 (1993)
25. Smith, J.M.: Chemical Engineering Kinetics. McGraw-Hill, Tokyo (1981)
26. Tardos, G.I., Gutfinger, C., Abuaf, N.: High Peclet mass transfer to a sphere in a fixed or fluidized bed. AIChE J. **22**, 1147–1150 (1976)
27. Valdes-Parada, F.J., Ochoa-Tapia, J.A., Alvarez-Ramirez, J.: Effective medium equations for fractional fick's law in porous media. Phys. A **373**, 339–353 (2007)
28. Whitaker, S.: Diffusion and dispersion in porous media. AIChE J. **13**, 420–427 (1967)
29. Wilson, E.J., Geankoplis, C.J.: Liquid mass transfer at very low reynolds numbers in packed beds. Ind. Eng. Chem. Fund **5**, 9–14 (1966)
30. Zanotti, F., Carbonell, R.G.: Development of transport equations for multiphase systems I: general development for two-phase systems. Chem. Eng. Sci. **39**, 263–278 (1984)
31. Zanotti, F., Carbonell, R.G.: Development of transport equations for multiphase systems II: application to one-dimensional axi-symmetric flows of two-phases. Chem. Eng. Sci. **39**, 279–297 (1984)

Chapter 6
Experimental and Numerical Investigation of Mass Transport in Porous Media

6.1 Measurement of Molecular Diffusion Coefficients

Mass transfer is a topic of central importance in chemical engineering, and molecular diffusion is the basic physical mechanism underlying mass transfer processes, even when convection comes into play. As a result, values of the molecular diffusion coefficient (D_m) are required for mass transfer calculations and extensive tabulations of this parameter have been prepared in the past (see [38, 41]). Equations for the prediction of D_m are also available and, in general, they have reasonable accuracy [44]. Nevertheless, it is important to determine values D_m experimentally, in many instances.

In the simple experiment described here, consideration is given to the process of mass transfer from a volatile solid sphere (a moth ball), buried in a packed bed of inert particles (sand or glass ballotini), through which air is forced to flow continuously. And also, to the similar process of dissolution of slightly soluble spheres buried in a packed beds of inerts through which water flows. These are important "model situations" for the understanding of such processes as char combustion in fluidised beds [42] and leaching of ore (or contaminant) from buried rocks (or buried waste).

When these processes are performed with very low fluid velocities, the rates of mass transfer are strongly determined by molecular diffusion and the experiment may be used to provide an accurate method for the measurement of the diffusion coefficient. It is an entirely novel method that has the added interest of easily providing data at temperatures and pressures that differ significantly from ambient values. This is not always easy to achieve with other methods [35].

As an introduction to the present analysis, it is useful to consider the simple situation depicted in Fig. 6.1a, where the vapour liberated by a buried sphere travels through the interstices of a packed bed of inerts, as a result of molecular diffusion.

F. A. Coutelieris and J. M. P. Q. Delgado, *Transport Processes in Porous Media*, Advanced Structured Materials 20, DOI: 10.1007/978-3-642-27910-2_6,

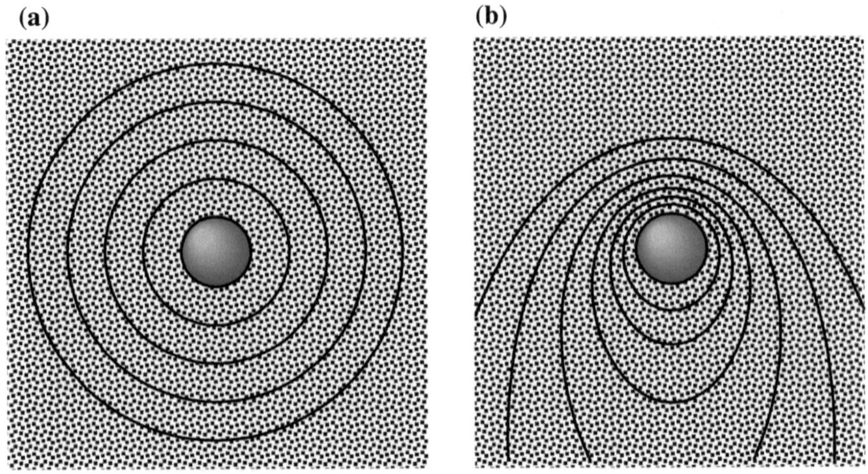

Fig. 6.1 Sketch of iso-concentration surfaces around active sphere of radius b for **a** pure diffusion, and **b** diffusion with convection

6.1.1 Diffusion Alone

The sphere of slightly volatile solid is assumed to be buried in a packed bed of sand, of "infinite extent", the interstices of the bed being filled with a stagnant gas (e.g. air) that is assumed to be free of solute at a large distance from the sphere (*i.e.* $C \to 0$ as $r \to \infty$). At $r = b$, where the active sphere contacts the gas phase, the molar concentration of solute in the gas is $C^* = P_V/RT$, where P_V is the equilibrium vapour pressure of the solute at the temperature (T) of the experiment.

The concentration gradient will give rise to a process of outwards diffusion of solute, but if diffusion is slow, the size of the sphere is taken to be constant, during one experiment; in other words, the assumption of a quasi-steady state is legitimate.

Under that assumption, the rate of diffusion across a spherical surface of radius r, concentric with the solid sphere, will be independent of r. From Fick's law,

$$n = -D'_{\mathrm{m}}\left(4\pi r^2 \varepsilon\right)\frac{\mathrm{d}C}{\mathrm{d}r} \tag{6.1}$$

where $D'_{\mathrm{m}} = D_{\mathrm{m}}/\tau$ is the effective diffusion coefficient (τ being the tortuosity factor, accepted to be $\sqrt{2}$ for packed beds of granular materials [48] and ε is the bed voidage (assumed constant throughout). Since n is independent of r, integration of Eq. 6.1, between the limits ($r = b, C = C^*$) and ($r \to \infty$, $C \to 0$), gives

$$n = 4D'_{\mathrm{m}}\varepsilon\,\pi\, b\,(C^* - 0) = 2\frac{D'_{\mathrm{m}}}{d_1}\varepsilon\,\pi\, d_1^2\,(C^* - 0) \tag{6.2}$$

where d_1 $(= 2b)$ is the diameter of the active sphere. Identifying $(\pi\, d_1^2)$ as the area of the active sphere, helps recognize the expression for the mass transfer coefficient as $k = 2\varepsilon D'_{\mathrm{m}}/d_1$. Introduction of the Sherwood number, $\mathrm{Sh}' = kd_1/D'_{\mathrm{m}}$, shows that for mass transfer by pure diffusion, around a buried sphere,

$$\frac{\mathrm{Sh}'}{\varepsilon} = 2 \tag{6.3}$$

an expression similar to the well known result, Sh = 2, for pure diffusion around a sphere in an unbounded fluid [48].

6.1.2 *Diffusion with Convection*

Figure 6.1b represents again the buried sphere, but now exposed to a constant flow of gas, with average interstitial velocity u_0. The solute still diffuses away from the surface of the sphere, but the rate of mass transfer is now enhanced, since the solute is continuously swept away by the moving fluid. The surfaces of equal concentration are no longer spheres. The analysis of the physical situation is complicated, but it has been worked out in detail [23]. For the conditions of interest in the present work, the expression

$$\frac{\mathrm{Sh}'}{\varepsilon} = \left[4 + \frac{4}{5}\left(\mathrm{Pe}'_{\mathrm{p}}\frac{d_1}{d}\right)^{2/3} + \frac{4}{\pi}\left(\mathrm{Pe}'_{\mathrm{p}}\frac{d_1}{d}\right)\right]^{1/2} \tag{6.4}$$

has been shown to give accurate values of Sh'/ε, where $\mathrm{Pe}'_{\mathrm{p}} = u_0 d/D'_{\mathrm{m}}$ is the Peclet number based on the diameter of the inert particles (*d*) making up the bed. It may be easily seen that Eq. 6.4 reduces to Eq. 6.3, in the limit of low Pe' ($= \mathrm{Pe}'_{\mathrm{p}} d_1/d$). As $\mathrm{Pe}'_{\mathrm{p}}$ is increased, it is known that Eq. 6.4 becomes inaccurate [23]. At $\mathrm{Pe}'_{\mathrm{p}} \cong 1$, Eq. 6.4 is found to be accurate to within 5%, and therefore this value of $\mathrm{Pe}'_{\mathrm{p}}$ should not be exceeded. This is because convective dispersion (rather than molecular diffusion) would then become the relevant mechanism of mass transfer in the packed bed. However, in the case of the experiments described below, conditions are strictly restricted to the range of applicability of Eq. 6.4.

The essence of the experimental method proposed relies on burying a weighed sphere of naphthalene (with initial mass m_0 and diameter d_1) in a packed bed of sand of known porosity (ε) and continuously forcing a metered stream of air (volumetric flowrate, v) through the packed bed (cross sectional area, A) for a given time interval, Δt. Weighing the sphere at the end of the time interval (mass m_{t}) gives the rate of sublimation as $(m_0 - m_{\mathrm{t}})/\Delta t$ and the mass transfer coefficient may be calculated from

$$k = \frac{m_0 - m_t}{\Delta t\, M\, \pi d_1^2\, (p_V/RT)} \tag{6.5}$$

where M is the molecular weight of naphthalene.

With the interstitial velocity given by $u_0 = v/(A\varepsilon)$, the only unknown in Eq. 6.4 is D'_m. A simple way of solving this equation relies on re-arranging it, through multiplication by D'_m ($= D_m/\sqrt{2}$), to obtain

$$\frac{kd_1}{\varepsilon} = D'_m \left[4 + \frac{4}{5}\left(\frac{u_0 d_1}{D'_m}\right)^{2/3} + \frac{4}{\pi}\left(\frac{u_0 d_1}{D'_m}\right)\right]^{1/2} \tag{6.6}$$

Several models can be found in the literature for the prediction of diffusion coefficients in binary systems [48]. Fuller et al. [21] suggest the use of the following equation

$$D_m = \frac{\left(\frac{M_A + M_B}{M_A M_B}\right)^{1/2} T^{1.75}}{P\left[(\Sigma v)_A^{1/3} + (\Sigma v)_B^{1/3}\right]^2} \times 10^{-7} \tag{6.7}$$

where P is the absolute pressure (in bar), M_A and M_B are, respectively, the molecular weights of components A and B (in g/mol), and $(\Sigma v)_A$ and $(\Sigma v)_B$ are the values of the atomic diffusion volumes of components A and B, which are tabulated in Reid et al. [44].

Equation 6.7 predicts a proportionality between D_m and $T^{1.75}/P$, and this is an important result that may be demonstrated using the experimental technique described in the following section.

6.1.3 Experiments

An example of the experimental setup is sketched in Fig. 6.2. The air supply may determine the maximum pressure at which the experiment can be performed. The tubing and valves connecting the air supply to the low pressure end of the rig will be standard material, for the maximum working pressure intended. The test column, containing the packed bed of sieved sand (or glass ballotini) may be made from a short piece of stainless steel tube (typically 80 mm internal diameter and 110 mm long) flanged at both ends. The test column is best kept vertical, to avoid settling of the packing to one side, and a downward gas flow will prevent unwanted fluidization of the bed material. A piece of some sort of gauze should be placed over the bottom plate of the test column to prevent the granular material from going into the tubing.

If the rotameter gives the flowrate v_R, at pressure P_R and temperature T_R, the actual flowrate in the test column is given as $v = v_R(P_R/P)(T/T_R)$, where P is the absolute pressure indicated in the manometer connected to the test column. The vapour pressure is calculated using Uno's correlation [53].

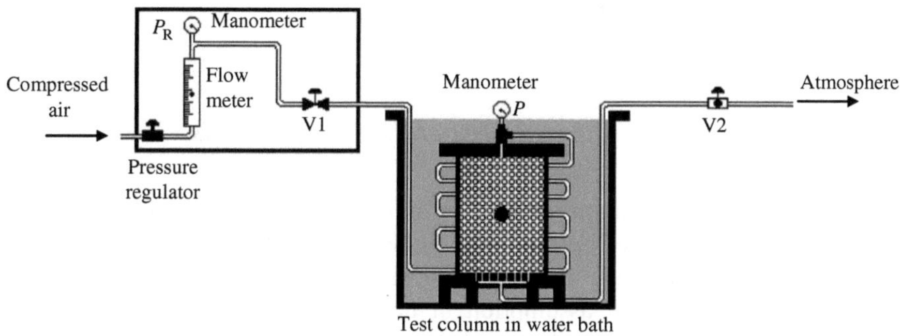

Fig. 6.2 Experimental set-up

In experiments at temperatures differing from ambient, the test column and a significant length of the tubing feeding it have to be kept in a constant temperature bath.

The naphthalene spheres used in the experiments were solidified from the melt, in simple moulds made of silicone rubber for the purpose. The diameter of each sphere was measured with callipers (along three perpendicular directions, to ensure near sphericity) and it was typically about 20 or 25 mm.

Each sphere was weighed accurately (initial mass, m_0), in an analytical balance, before burying it (with some care, to avoid unwanted erosion) near the middle of the packed bed, which had been previously immersed in the constant temperature bath for a long enough period of time. The top plate of the test column was then bolted in place and the column was tapped gently, a few times, to ensure close packing of the granular material. The test column and associated tubing were then immersed again in the constant temperature bath and the air supply was then connected, to give the intended test pressure and gas flowrate. The air flowrate was kept constant for a time Δt, following which it was interrupted and the test column was opened, to remove the sphere for weighing (final mass, m_t).

Each experiment lasted between 50 min and 60 h, the time being chosen to give a measurable loss of weight (typically, 0.05 g), that nevertheless would not correspond to a significant variation in the diameter of the sphere. The time intervals were however sufficiently long to reduce inaccuracies due to evaporation in the stages of sphere introduction and removal from the test column.

The experimental measurements of mass transfer in water were performed on the dissolution of individual spheres of for example, 2-napthol or benzoic acid, buried in beds of sand through which a metered stream of distilled water (which had been previously deaerated under vacuum) was forced to flow steadily.

Instead of weighing the sphere, the concentration of solute in the outlet stream, C_{out}, was continuously measured to give the rate of dissolution. C_{out} was measured by means of a UV/VIS spectrophotometer. When steady state was reached, the rate of dissolution of the solid could be found directly from $n = QC_{\text{out}}$, where Q is the measured volumetric flow rate of water.

6.2 Measurement of Dispersion Coefficients (Axial and Radial)

6.2.1 *Measurement of Axial Dispersion Coefficients*

Since the development of the dispersion approximation for the study of solute transport in capillary tubes by Taylor [52], the flow of the tracer is described by dispersion due to molecular diffusion and radial velocity variations. In packed beds, with $D/d > 15$, the assumption of flat velocity profiles and porosity is reasonable as point out by Akehata and Sato [1] and Gunn [25] and later showed by the experimental studies of Stephenson and Stewart [50] and Gunn and Pryce [26], that suggested $D/d > 10$.

Imagine a packed bed of uniform porosity (ε), contained in a long column of length L along which liquid flows at a superficial velocity U (the interstitial velocity is then $u = U/\varepsilon$) and initial concentration of solute C_0, in which a tracer with continuously injection and concentration of soluteC_S, is dispersed in radial and axial direction. Taking a small control volume inside this boundary layer, a material balance on the solute, with length dz and width dr, leads to (see [25])

$$D_L \frac{\partial^2 C}{\partial z^2} + \frac{1}{r}\frac{\partial}{\partial r}\left(D_T r \frac{\partial C}{\partial r}\right) - u\frac{\partial C}{\partial z} = \frac{\partial C}{\partial t} \tag{6.8}$$

Klinkenberg et al. [33] and Bruinzeel et al. [9] show that radial dispersion can be neglected in comparison with axial dispersion for a small ratio of column diameter to length (D/L) and large fluid velocity. The partial differential equation describing tracer transport in the bed reduces then to

$$D_L \frac{\partial^2 C}{\partial z^2} - u\frac{\partial C}{\partial z} = \frac{\partial C}{\partial t} \tag{6.9}$$

where z measures length along the bed, and if L is sufficiently large (semi-infinite bed) the appropriate boundary conditions are

$$C = C_0 \qquad 0 \leq z \leq L \qquad t = 0 \tag{6.10a}$$

$$uC_S = uC - D_L \frac{\partial C}{\partial z} \qquad z = 0 \qquad t > 0 \tag{6.10b}$$

$$\frac{\partial C}{\partial z} = 0 \qquad z = L \qquad t > 0 \tag{6.10c}$$

For a step input, the concentration at the outlet of the bed ($z = L$) can be obtained by Carslaw and Jaeger [10], who give the exact solution of the equivalent heat transfer problem. However, a study developed by Harrison et al. [29] showed that the boundary conditions developed by Danckwerts [15], for an infinite system, hold adequately for a finite system provided $uL/D_L \geq 10$, So, for a step input (from C_0 to C_S), the concentration at the outlet of the bed ($z = L$) is known [15] to be given, if L is sufficiently large, by

$$\mathrm{F}(\theta) = \frac{1}{2}\left[1 - \operatorname{erf}\left(\sqrt{\frac{L\mathrm{Pe_L}}{\theta d}}\frac{(1-\theta)}{2}\right)\right] \tag{6.11}$$

or for a pulse response by

$$E(\theta) = \frac{1}{2}\left(\frac{L\mathrm{Pe_L}}{\pi\theta d}\right)^{1/2} \times \exp\left[\frac{-L\mathrm{Pe_L}(1-\theta)^2}{4\theta d}\right] \tag{6.12}$$

Rifai et al. [45] and Ogata and Banks [39] showed that the solution of Eq. 6.9 with the boundary conditions and initial condition given by Eqs. 6.10a–c is

$$\begin{aligned}\mathrm{F}(\theta) =& \frac{1}{2}\left[1 - \operatorname{erf}\left(\sqrt{\frac{L\mathrm{Pe_L}}{\theta d}}\frac{(1-\theta)}{2}\right)\right]\\ &+\frac{1}{2}\left[1 - \operatorname{erf}\left(\sqrt{\frac{L\mathrm{Pe_L}}{\theta d}}\frac{(1+\theta)}{2}\right)\right]\exp\left(\frac{L}{d}\mathrm{Pe_L}\right)\end{aligned} \tag{6.13}$$

However, Ogata and Banks [39] showed that for large molecular Peclet numbers (say, $uL/D_L > 100$), the advection dominates and the second term in the right-hand side can be neglected, with an error lesser than 5%, and Eq. 6.13 reduces to Eq. 6.11.

6.2.1.1 Experiments

Typically, dispersion along the direction of flow is studied by following the distortion of some concentration wave (of a tracer), as it progresses along the packing. In our experiments, a step in tracer concentration was introduced at the top of a long packed bed, of constant cross section, and the variation of tracer concentration, continuously in the stream leaving the bed, was recorded.

The partial differential equation describing tracer transport in the bed is given by Eq. 6.9 and for a step input (from C_0 to C_S), the concentration at the outlet of the bed ($z = L$) is known [15] to be given by

$$\frac{C - C_0}{C_S - C_0} = \frac{1}{2}\left[1 - \operatorname{erf}\left(\frac{L - ut}{2\sqrt{D_L t}}\right)\right] \tag{6.14}$$

if L is sufficiently large. This result may be written as Eq. 6.11, where $\mathrm{F} = (C - C_0)/(C_S - C_0)$ is the dimensionless concentration rise and $\boldsymbol{\theta} = t/\bar{t}$ is the dimensionless time ($\bar{t}$ being the mean residence time of fluid in the bed).

For each set of C *vs.* t values, obtained in one experiment, the values of F *vs.* θ were calculated and the value of $\mathrm{Pe_L}$ determined to give the least deviation

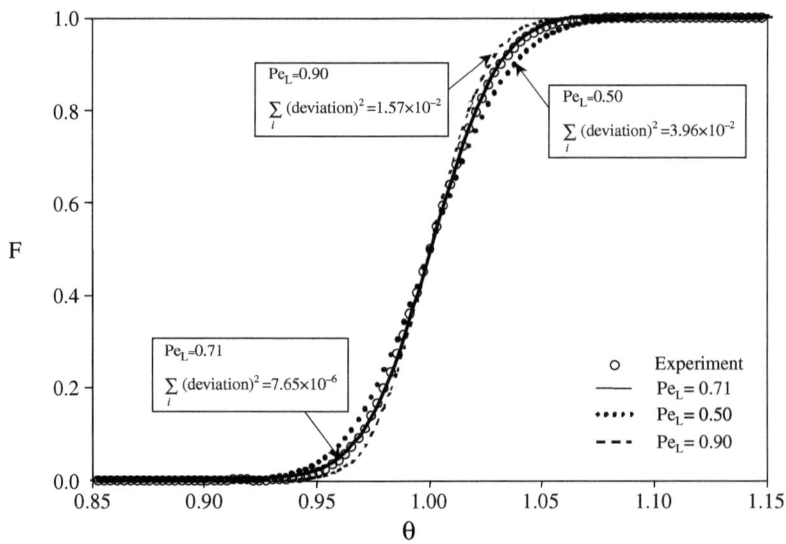

Fig. 6.3 Comparison of experiment with Eq. 6.4 for three values of Pe_L

between the experimental points and the line representing Eq. 6.11. Figure 6.3 helps illustrate the method and it also gives some idea about its sensitivity to variations inPe_L.

The rig used in the majority of our experiments (including all those at temperature above ambient) is sketched in Fig. 6.4, but the various items are not to scale. Two large stainless steel reservoirs (about 0.3 m in diameter and 1.0 m tall) were used to keep the distilled water and the dilute solution of sodium chloride (up to 1.5 kg/m^3 in salt) immersed in a silicone oil thermostatic bath. The use of very dilute salt solutions and the care taken in equalizing the temperature of both liquids is needed to avoid dispersion by natural convection. Both the distilled water and the salt solution were degassed "in situ", by bubbling under vacuum, to avoid liberation of small air bubbles inside the test column, at the higher temperatures. Both reservoirs were connected (at the top) to the compressed air line (approx.: 4 bar) to have the rig permanently pressurised and help discharge the liquid through the flow regulating valve V, open to the atmosphere.

At the top of the test column (3.0 m long and 0.047 m in diameter, made of brass), a three-way valve (T) could be manipulated to select the feed. At the beginning of each experiment it was turned to let distilled water flow through the bed, until no salt was detected in the conductivity meter monitoring the exit stream. After that, it was turned to allow the salt solution to go through the column, at a constant flowrate, measured by an orifice meter connected to a differential pressure transducer. The dead space between the three-way valve and the top of the bed was less than 0.1% of the void space in the bed proper and this is important to ensure that a sharp step input of tracer is obtained in the bed. The conductivity

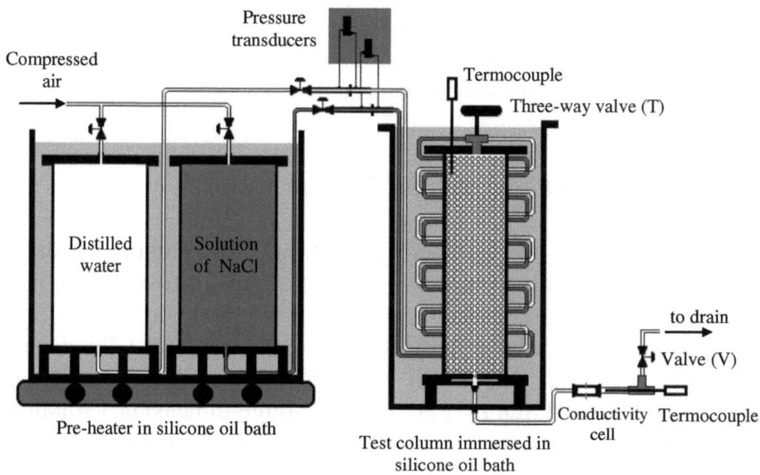

Fig. 6.4 Diagram of experimental set-up

cell at the exit of the test column was connected to a micro-computer and values of the salt concentration were recorded at a frequency of 10 Hz. Care was taken to have a negligible volume of liquid between the bottom of the bed and the conductivity cell in order to avoid distortion of the signal read by the cell. A typical record of outlet tracer concentration is shown in Fig. 6.3 and from the sensitivity of the curves representing Eq. 6.11, to the value of Pe_L, it may be guessed that the values of D_L obtained are accurate to within $\pm 15\%$.

6.2.2 Measurement of Radial Dispersion Coefficients

The radial dispersion coefficient can be determined by plotting (% composition : C_{10} and C_{90}) *vs* (distance from 50% composition) on arithmetic-probability paper [40]. The dispersion coefficient can be calculated by

$$D_T = \frac{u}{L}\left(\frac{C_{90} - C_{10}}{3.625}\right)^2 \tag{6.15}$$

The most widely used techniques for the measurement of lateral dispersion are the continuous point source and the instantaneous finite source methods (see *e.g.* [46]), which rely on the injection of tracer in a flowing liquid, followed by tracer detection at several points, downstream of the injection point. If at time $t = 0$ a tracer is injected into the porous medium from an injector, for the continuous point source method the tip of the injector is taken as the tracer origin. For the instantaneous finite source method the origin lays just down-gradient of the tracer injector.

Several authors like Roemer et al. [47] and Gunn and Pryce [26] used the solution of Eq. 6.15 when the axial dispersion coefficient is taken equal to radial dispersion coefficient. However, in this work we only consider experimental techniques where axial dispersion is neglected.

6.2.2.1 "Instantaneous Finite Source" Method

The method adopted by some authors like Dorweiler and Fahien [16] and Fahien and Smith [19] is based that the tracer is fed into the main stream at a point on the axis on the column.

The analytical model for an instantaneous finite source in one dimension is first presented by Crank [13]. Baetsle [6] extended the model to three-dimensional dispersion. Hunt [32] and Sun [51] provided the three-dimensional solution to the advection–dispersion equation [6] using different mathematical analysis. Van Genuchten and Alves [54] presented a number of analytical solutions of the one-dimensional convective–dispersive solute transport equation.

Tracer concentration should be low enough to avoid density-induced flow effects. The tracer should be conserved (*i.e.* not destroyed) in the experiment and the distribution of flow rates at the outlet must be the same as in the feed so as not to induce complications in the flow field.

Radial dispersion may be evaluated by injecting a steady flow of a tracer at a point of a test section column. For a boundary layer, which is thin in comparison with the length of the axial distance (L), axial dispersion will be negligible. Taking a radial co-ordinate, r, to measure distance to the axis of the bed and a co-ordinate z, to measure distance along the average flow direction, the differential mass balance on the solute reads

$$\frac{D_{\mathrm{T}}}{r}\frac{\partial}{\partial r}\left(r\frac{\partial C}{\partial r}\right) = u\frac{\partial C}{\partial Z} \tag{6.16}$$

where D_{T} is the radial dispersion coefficient. Fahien and Smith [19] solved the differential dispersion Eq. 6.16 with

$$z = 0 \quad 0<r<R_{\mathrm{i}} \quad C = C_0 \tag{6.17a}$$

$$z = 0 \quad R_{\mathrm{i}}<r<R \quad C = 0 \tag{6.17b}$$

$$\text{all } z \quad r = R;\ r = R_{\mathrm{i}} \quad \frac{\partial C}{\partial r} = 0 \tag{6.17c}$$

and the solution of Eq. 6.16 with the boundary conditions (6.17a–c) is

$$\frac{C}{C_0} = 1 + \frac{2R}{R_{\mathrm{i}}}\sum_{\mathrm{n}=1}^{\infty}\frac{\mathrm{J}_1(\beta_{\mathrm{n}}R_{\mathrm{i}}/R)\mathrm{J}_0(\beta_{\mathrm{n}}r/R)}{\beta_{\mathrm{n}}\mathrm{J}_0^2(\beta_{\mathrm{n}})}\frac{z}{R}\exp\left[-\frac{\beta_{\mathrm{n}}^2}{\mathrm{Pe}_{\mathrm{T}}}\frac{z}{R}\right] \tag{6.18}$$

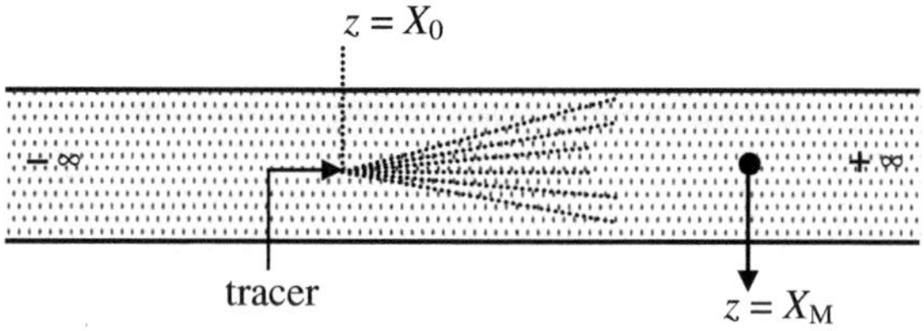

Fig. 6.5 A schematical diagram of test section for radial dispersion

where J_0 and J_1 are the Bessel functions of the first kind, of order 0 and 1, respectively, and the β_n are the positive roots of the Bessel function of the first kind, of order 1.

6.2.2.2 "Continuous Point Source" Method

This method is based on the measurement of radial mass exchange between two coaxial portions of a packed bed, along which liquid flows, parallel to the axis; the feed to the central portion is water containing a small amount of sodium chloride and that to the outer portion is pure water.

Klinkenberg et al. [33] derived an analytical solution for Eq. 6.16, neglecting the effect of injector radius (see Fig. 6.5), with the boundary conditions given by

$$z = +\infty \quad \text{all } R \quad C = C_0 \tag{6.19a}$$

$$z = -\infty \quad \text{all } R \quad C = 0 \tag{6.19b}$$

$$\text{all } z \quad r = R;\ r = 0 \quad \frac{\partial C}{\partial r} = 0 \tag{6.19c}$$

and the solution of Eq. 6.16 with the boundary conditions (6.19a–6.19c) is

$$\frac{C}{C_0} = 1 + \sum_{n=1}^{\infty} \frac{J_0(\beta_n r/R)}{J_0^2(\beta_n)} \exp\left[-\frac{\beta_n^2 D_T z}{R^2 u}\right] \tag{6.20}$$

where J_0 is the Bessel function of the first kind, of order 0, and the β_n are the positive roots of the Bessel function of the first kind, of order 1.

Plautz and Johnstone [43] and Sinclair and Potter [49] used Eq. 6.9 for an infinite case, where no boundary is present, of mass diffusion from a point source. The result with axial dispersion neglected was given by Carslaw and Jaeger [10]

$$\frac{C}{C_0} = \frac{R^2 u}{4D_T z} \exp\left(-\frac{r^2 u}{4D_T z}\right) \tag{6.21}$$

This solution includes a simplification possible when $z/r > 5$ (axial dispersion neglected).

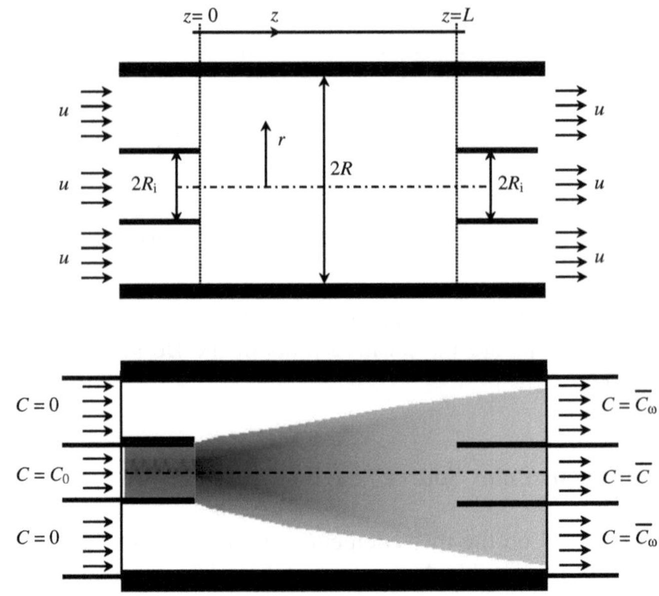

Fig. 6.6 Sketch of boundary conditions proposed by [30]

Blackwell [8] assumes the effect of radius injector and gives the analytical solution to the differential equation describing radial dispersion in the absence of axial dispersion. Hiby and Schummer [30] presented a solution of Eq. 6.16 that considered the tracer pipe to be of significant diameter compared to the diameter of the bed (see Fig. 6.6), and the boundary conditions adopted were

$$\mathrm{z} > 0 \quad r = R \quad \frac{\partial C}{\partial r} = 0 \tag{6.22a}$$

$$\mathrm{z} = 0 \quad r < R_\mathrm{i} \quad C = C_0 \tag{6.22b}$$

$$\mathrm{z} = 0 \quad R_\mathrm{i} < r < R \quad C = 0 \tag{6.22c}$$

On the assumption that D_T and u are independent of position, the solution of Eq. 6.9 following Hiby and Schummer [30] gives, for the resulting outlet average concentration in the inner stream of liquid,

$$\frac{\bar{C}}{C_0} = 4\sum_{\mathrm{n}=0}^{\infty} \frac{\mathrm{J}_1^2(\beta_\mathrm{n} R_\mathrm{i}/R)}{\beta_\mathrm{n}^2 \mathrm{J}_0^2(\beta_\mathrm{n})} \exp\left[-\frac{Ld}{\mathrm{Pe}_\mathrm{T}}\left(\frac{\beta_\mathrm{n}}{R}\right)^2\right] \tag{6.23}$$

where J_0 and J_1 are the Bessel function of the first kind, of orders 0 and 1, respectively, and the β_n are the positive roots of the Bessel functions of the first kind, of order 1. The measurement of $\bar{C}$ and C_0 provides a method for the determination of Pe_T (and therefore of D_T), since all other parameters in the equation are known.

Harleman and Rumer [28] and Han et al. [27] consider a steady-state experiment in a rectangular column. The authors solved the differential equation with the boundary conditions,

$$C = C_0 \quad x = 0 \quad 0 < y < +\infty \tag{6.24a}$$

$$C = 0 \quad x = 0 \quad -\infty < y < 0 \tag{6.24b}$$

$$\frac{\partial C}{\partial y} = 0 \quad \text{all } x \quad y \to \pm\infty \tag{6.24c}$$

and the solution obtained for a step input in concentration, is

$$\frac{C}{C_0} = \frac{1}{2}\left[1 - \mathrm{erf}\left(\sqrt{\frac{\mathrm{Pe_T}}{Ld}}\frac{y}{2}\right)\right] \tag{6.25}$$

6.2.2.3 Mass Transfer From a Flat Surface Aligned With the Flow

Coelho and Guedes de Carvalho [12] developed a new experimental technique, based on the measurement of the rate of dissolution of planar or cylindrical surfaces, buried in the bed of inert particles and aligned with the flow direction.

Figure 6.7a sketches a section through a packed bed along which liquid is flowing, close to a flat wall, part of which ($0 < x < L$) is slightly soluble. Liquid flow will be taken to be steady, with uniform average interstitial velocity u, and if the concentration of solute in the liquid fed to the bed is C_0 and the solubility of the solid in the wall is C^*, a mass transfer boundary layer will develop, across which the solute concentration drops from $C = C^*$, at $y = 0$, to $C \to C_0$, for large y.

The question of how large is meant by a "large y" needs some clarification. Obviously, if L were only of the order of a few particle diameters, and u were large, the concentration of solute would fall to C_0 over a distance of less than one particle diameter. In that case, flow in the bulk of the packed bed would have little influence on the mass transfer process, which would be dominated by diffusion in a thin layer of liquid, adjacent to the soluble surface. Already for large L and low u, the thickness of the mass transfer boundary layer will grow from zero, at $x = 0$, to a value of several particle diameters, at $x = L$ and the process of mass transfer will then be determined by a competition between advection and dispersion in the bulk of the bed. Now, it is well known (*e.g.* [55]) that the voidage of a packed bed (and therefore the fluid velocity) is higher near a containing flat wall, but in the case of Guedes de Carvalho and Delgado [24] experiments it may be considered that such a non-uniformity will have negligible effect. For one thing, we work with bed particles of between 0.2 and 0.5 mm and therefore the region of increased voidage will be very thin. Furthermore, because the inert particles making up the bed indent the soluble surface slightly, as dissolution takes place (and this slight indentation is easily confirmed when the piece of soluble solid is removed from

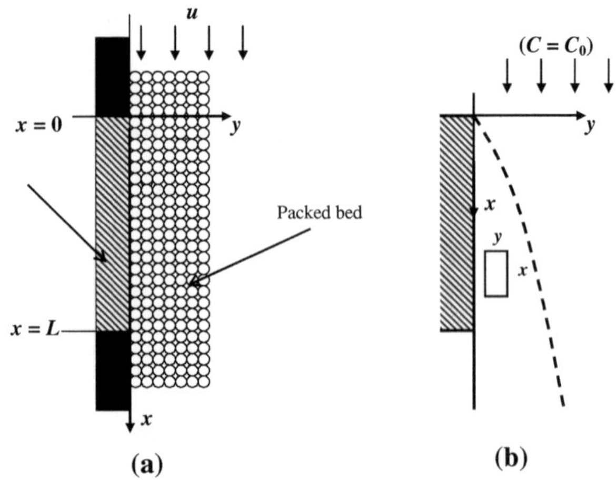

Fig. 6.7 **a** Flow along soluble slab, **b** Mass transfer boundary layer

within the bed), there is in fact virtually no near wall region of higher voidage. Confirmation of these assumptions is given by the results of the experiments described below.

Taking a small control volume inside this boundary layer (see Fig. 6.7b), with side lengths $\delta x,\ \delta y$ and unity (perpendicular to the figure), it is possible to perform a mass balance on the solute, for the steady state. If the boundary layer is thin, compared to the length of the soluble slab, axial dispersion is likely to be negligible, since the surface $y = 0, 0 < x < L$, is a surface of constant concentration ($C = C^*$).

Noting that the surface $y = 0,\ 0 < x < L$, is a surface of constant concentration, along which $\partial^2 C/\partial x^2 = 0$ and axial dispersion will be negligible, for a boundary layer which is thin in comparison with the length of the soluble slab. (A conservative criterion for this approximation to be valid is $L/d > 20$). For a slab the equation of diffusion in one dimension is

$$u\frac{\partial C}{\partial x} = D_{\mathrm{T}}\frac{\partial^2 C}{\partial y^2} \tag{6.26}$$

to be solved with

$$C = C_0 \quad x = 0 \quad y > 0 \tag{6.27a}$$

$$C = C^* \quad x > 0 \quad y = 0 \tag{6.27b}$$

$$C \to C_0 \quad x > 0 \quad y \to \infty \tag{6.27c}$$

The solution is

$$\frac{C - C_0}{C^* - C_0} = \operatorname{erfc}\left(\frac{y}{2\sqrt{D_T x/u}}\right) \tag{6.28}$$

and the flux of dissolution at any point on the slab surface may be obtained from (6.28) as

$$N = -D_T \varepsilon \left(\frac{\partial C}{\partial y}\right)_{y=0} = (C^* - C_0)\varepsilon \left(\frac{D_T}{\pi x/u}\right)^{1/2} \quad (6.29)$$

The instant rate of solid dissolution over the whole slab surface may now be calculated by integration of the local flux; taking a width b along the surface of the solid, perpendicular to the flow direction, there results

$$n = \int_0^L N\, b\, dx = (C^* - C_0)\varepsilon\, b\, L \left(\frac{4D_\mathrm{T}}{\pi L/u}\right)^{1/2} \quad (6.30)$$

and it is useful to define the coefficient

$$k = \frac{n}{(bL)(C^* - C_0)} = \varepsilon \left(\frac{4D_\mathrm{T}}{\pi L/u}\right)^{1/2} \quad (6.31)$$

This result shows how the measurement of the rate of dissolution of the solid, which is directly related to the average mass transfer coefficient, may be used to determine the coefficient of radial dispersion in the bed.

A simple way of checking the result in Eq. 6.31 is afforded by the predicted proportionality between k and the inverse square root of L. Experiments performed by Coelho and Guedes de Carvalho [12] with a wide range of slab lengths, both for the dissolution of benzoic acid in water and the sublimation of naphthalene in air, confirm the general validity of the above theory, provided that the approximate criterion

$$\frac{L}{d} \geq 0.62 \left(\frac{ud}{D_m}\right) \quad (6.32)$$

is observed, where D_m is the molecular diffusion coefficient of the solute. When the above criterion is not observed, the near wall film resistance to diffusion will have to be taken into account and approximate ways of doing this are described by Coelho and Guedes de Carvalho [12].

The similarity between the result given by Eq. 6.32 and that obtained by Higbie [31], for gas–liquid mass transfer by surface renewal, is striking. Equation 6.24a–c simply states that the average mass transfer coefficient, for the soluble wall, is that corresponding to surface renewal with a time of contact $t_\mathrm{c} = L/u$ and an apparent diffusion coefficient D_T.

6.2.2.4 Mass Transfer From a Cylinder Aligned With the Flow

For practical reasons, it proves simpler to perform experiments in which the dissolving solid is a cylinder, aligned with the flow direction and it is important to know the theoretical expressions relating the average mass transfer coefficient with the coefficient of dispersion, D_{T}, for that situation.

Fortunately, under appropriate conditions, easy to reproduce in the laboratory, the thickness of the mass transfer boundary layer is small in comparison with the radius of the dissolving cylinder and under such circumstances, the analysis presented above, for dissolution from a flat surface, is still applicable with good accuracy.

However, there are instances in which this simplification is not valid and an exact solution may be worked out in cylindrical co-ordinates, as shown by Coelho and Guedes de Carvalho [12].

The resulting expression for k is cumbersome to evaluate, but for small values of the parameter $\theta_c = D_T t_c / a^2$, where $t_c = L/u$ is the time of contact between liquid and solid, a good approximation is

$$k = \varepsilon \left(\frac{4D_{\mathrm{T}}}{\pi t_c}\right)^{1/2} \left(1 + \frac{\sqrt{\pi}}{4}\theta_c^{1/2} - \frac{1}{12}\theta_c + \frac{\sqrt{\pi}}{32}\theta_c^{3/2} - \cdots\right) \tag{6.33}$$

For higher values of θ_c, up to θ_c =0.4, the first four terms may be used, instead of the infinite series on the right hand side of Eq. 6.33, with an error of less than 1% in k.

6.2.2.5 Experiments

Experiments were performed on the dissolution of individual cylinders of 2-napthol, buried in beds of sand through which a metered stream of distilled water (which had been previously deaerated under vacuum) was forced to flow steadily, as sketched in Fig. 6.8. The beds of sand were contained in a stainless steel column 500 mm long and 100 mm in internal diameter, with the cylinder of 2-napthol placed co-axially inside it, all in vertical alignment. The cylinders of 2-napthol had a diameter of 20 mm and a length of 250 mm and they were tightly mounted in alignment between two rods of stainless steel (each 20 mm in diameter and 100 mm long). These two metal rods fulfilled the double purpose of covering the top and bottom of the 2-napthol cylinder and straightening the flow field, upstream and downstream from it; they also provided the points of support for alignment with the stainless steel column. Near the bottom of the stainless steel column, a perforated plate, covered with fine stainless steel wire mesh, was used to support the bed of sand and the cylinder inside it. The preparation and assemblage of the cylinders of 2-napthol followed closely the method detailed by Coelho and Guedes de Carvalho [12] for the preparation of cylinders of benzoic acid. The use of 2-napthol instead of benzoic acid was determined by the need to work at high temperatures and the relevant properties of 2-naphtol.

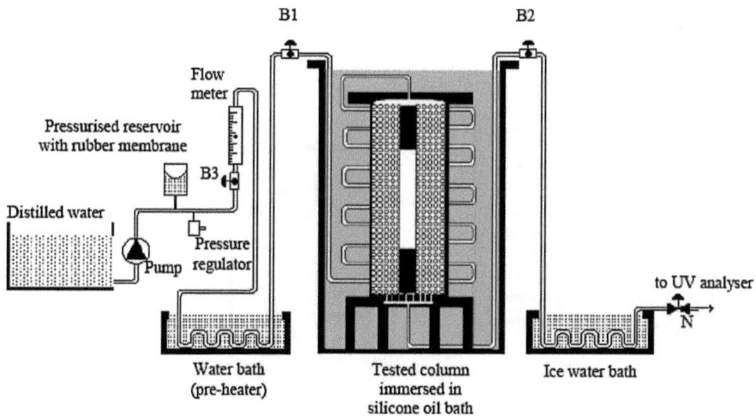

Fig. 6.8 Diagram of experimental set-up

Before any new series of runs, the cylinder buried in the sand (and when required, the sand itself) had to be replaced. In order to do that, ball valves B1 and B2 were closed and disconnected from the upstream and downstream piping, respectively, so as to allow the test column and the associated copper coil (and valves B1 and B2 in the ends) to be lifted from the thermostatic bath. Following that, the free ends of valves B1 and B2 were connected to plastic tubing that could be supplied with distilled water or directed to the drain. After removing the lid of the test column, distilled water was forced up through the bed of sand, so as to fluidise it slightly and allow the cylinder to be replaced without difficulty. As the water flow was stopped, to allow the sand particles to settle, the column was vibrated for a few seconds to give a good compaction of the bed. The lid of the test column was placed back in position and the stainless steel column (with attached copper coil and valves B1 and B2) was again immersed in the silicone oil bath. After connection to the main water circuit, valves B1, B2, and valve B3 were fully open and the flowrate of water fed to the top of the column was adjusted by means of valve N. The silicone oil bath was then heated to the required temperature of operation. A pre-heater helped warm up the water feed to near the temperature of the bath and an ice water bath was used to cool the liquid stream leaving the test column, before it passed the valve regulating the flow. In this way it was possible to perform experiments at temperatures up to the normal boiling point of water (and above it, if necessary).

The concentration of 2-naphtol in the outlet stream, C_{out}, was continuously measured by means of a UV/VIS Spectrophotometer, set at 274 nm, and when steady state was reached, the rate of dissolution of the solid could be found from $n = QC_{out}$, where Q is the measured volumetric flow rate of water.

For the second example, a rig was built as sketched in Fig. 6.9, that allowed experiments to be performed over a wide range of temperatures. Two large reservoirs, of about 0.06 m^3 internal volume, were used to store distilled water and a

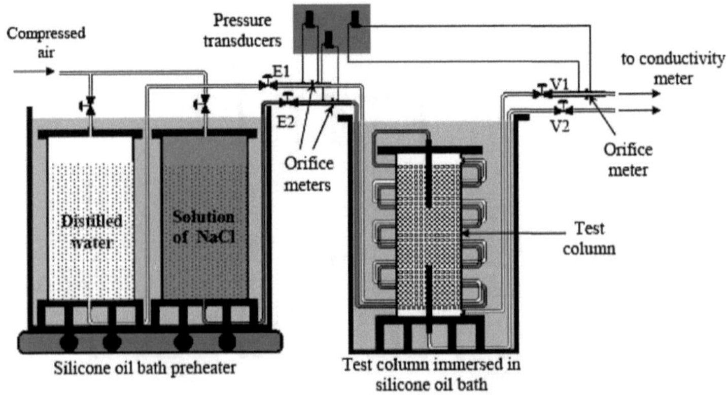

Fig. 6.9 Sketch of experimental set-up for measurement of radial dispersion coefficient

dilute solution of NaCl in distilled water with a concentration of salt of approximately 1.5 kg/m^3; both reservoirs were connected, at the top, to a source of air at 4 bar. The two reservoirs were kept in a thermostatic bath that was set at the temperature of the experiment, or at 353 K, when the experiments were performed above that temperature. The liquid held in each reservoir had been initially deaerated by stripping under vacuum, at room temperature, to avoid subsequent liberation of gas bubbles in the liquid streams going through the packed bed. In the test column, care was taken to have a good alignment of the axes of the inlet and outlet tubes of the core stream with that of the packed bed. The apparatus was kept in vertical alignment to avoid "settling" of the packing to any one side. Liquid flow was driven by the difference in pressure between the reservoirs and the exit sections, at atmospheric pressure. Valves V1 and V2 were used to adjust the outlet flowrate of each stream, to give equal velocity in the core tube and in the annulus, at the outlet; by acting slightly on valves E1 or E2 it was then possible to also impose equal velocities in the inlet streams to the bed. At the higher temperatures, the outlet streams had to be cooled before reaching valves V1 and V2, to avoid vaporization downstream of these valves, due to de-pressurization. The conductivity meter cell was built on a 30 mm nylon rod, 75 mm long, having a 10 mm hole drilled along its axis. Two platinum wires, 0.5 mm in diameter, crossed the wall of this nylon tube along opposite ends of one diameter of the mid cross section and they were glued in place, leaving a distance of about 8 mm between their tips. The platinum electrodes were connected to three 1 kΩ resistances (to within 0.01%) to form a Wheatstone bridge that was connected to a 9.30 V source, so that the liquid flowing through the cell acted as the variable resistance. The output of this conductivity meter was connected to a micro-computer and care was taken to calibrate the meter at the operating temperatures, whenever fresh solutions were introduced in the reservoirs. After allowing steady state to be reached, the rate of solute transfer from the core stream to the annulus is given simply as $n = v\left(C_0 - \overline{C}\right)$, where v is the volumetric flowrate of the core stream and $\overline{C}$ is the average concentration of salt in that stream, at the exit. Care was taken to check that the value of n agreed with $n = \omega \overline{C}_\omega$, where ω is the volumetric flowrate of the stream going through the annulus and $\overline{C}_\omega$ the corresponding average salt concentration, at the outlet.

6.3 Measurement of Solubility at Different Temperatures

Solubility is perhaps the most fundamental of all chemical phenomena. The significance of what dissolves what, to what extent, at what temperature and pressure, and the effects of other species, was recognized at a very early stage. In more recent times the importance of solubility phenomena has been acknowledged throughout science. For example, in the environment, solubility phenomena influence the weathering of rocks, the creation of soils, the composition of natural water bodies and the behaviour and fate of many chemicals.

The characteristic ability of water to behave as a polar solvent changes when water is subjected to high temperatures and pressures. As water becomes hotter, its molecules seem much more likely to interact with non-polar molecules. For example, at 300°C (and high pressure) water has dissolving properties very similar to acetone, a common organic solvent.

Also, solid–liquid and solid–gas mass transfer investigations with Newtonian or non-Newtonian fluids are frequently made by following the rate of dissolution of a low solubility solute. In all researches, accurate solubility data are required.

On mass transfer investigations in porous media, as in studies of dispersion coefficients and solute transport, the most common solutes used are benzoic acid, 2-naphthol, naphthalene, salicylic acid and succinic acid with water or air (see [56]). In these experiments, knowledge of accurate solubility data at different temperatures is very important, *i.e.*, for low solubility solutes.

The experiment proposed is simple and inexpensive, and it provides an accurate method for the measurement of solubilities of solid solutes in liquids and gases. Consider a vertical column of length L, containing a packed bed of soluble spherical particles of diameter d_1. If liquid flow is steady, with a uniform volumetric flowrate Q, if the concentration of solute in the liquid fed to the bed is c_0 and the solubility of the solid particle is c^*, a mass transfer boundary layer will develop.

In the analysis of results of experiments of dissolution of soluble spherical particles in liquid flow, the equation for dissolution rate is given by,

$$Q\frac{\partial c}{\partial x} = kS_{\mathrm{L}}(c^* - c_0) \tag{6.34}$$

where S_{L} is the active surface area per unit length and k is the average mass transfer coefficient. Given constant flowrate, uniformly distributed particles and isothermal conditions, Eq. 6.34 is integrated between the inlet and outlet conditions of the bed, $x = 0$ to $x = L$ and $c = c_0$ to $c = c$. The following equation results,

$$\frac{c - c_0}{c^* - c_0} = 1 - \exp\left(-\frac{kS_L}{Q}L\right) \tag{6.35}$$

In order to guarantee that the outlet stream is saturated, it's important to observe the approximate criterion $(c - c_0)/(c^* - c_0) > 0.999$ (error less that 0.1%). The number of soluble spheres presented in a packed bed is given by,

$$n = \frac{V_{\text{colunn}}}{V_{\text{particle}}} = \frac{3}{2}\frac{(1-\varepsilon)D^2L}{d_1^3} \tag{6.36}$$

and the general validity of the above theory holds, provided that the approximate criterion

$$\frac{6(1 - \varepsilon)kL}{u_0\varepsilon d_1} > 6.908 \tag{6.37}$$

If the criterion of Eq. 6.37 is to be satisfied, it is important to know the value of the average mass transfer coefficient, k, so as to be able to estimate the interstitial velocity of liquid, u_0.

6.3.1 Mass Transfer Around a Buried Soluble Sphere

For the propose of analysis, let as consider the situation of a slightly soluble sphere of diameter $d_1(=2a)$ buried in a bed of inert particles of diameter d (with $d \ll d_1$), packed uniformly (void fraction ε) around the spheres. The packed bed is assumed to be "infinite" in extent and a uniform interstitial velocity of liquid, u_0, is imposed, at a large distance from the spheres.

In order to obtain the flow field in the vicinity of the buried sphere, Darcy's law, $\mathrm{u} = -K\,\mathbf{grad}\,p$, is coupled with the continuity equation, $div\ \mathrm{u} = 0$, and Laplace's equation, $\nabla^2\phi = 0$, is obtained for the flow potential $\phi = kp$.

In terms of spherical coordinates (r, θ), the potential and stream functions are, respectively (see [14]),

$$\phi = -u_0\left[1 + \frac{1}{2}\left(\frac{a}{r}\right)^3\right] r\cos\theta \tag{6.38}$$

$$\psi = \frac{u_0}{2}\left[1 - \left(\frac{a}{r}\right)^3\right] r^2\sin^2\theta \tag{6.39}$$

and the velocity components are

$$u_r = \frac{\partial\phi}{\partial r} = -u_0\cos\theta\left[1 - \left(\frac{a}{r}\right)^3\right] \tag{6.40}$$

$$u_\theta = \frac{1}{r}\frac{\partial \phi}{\partial \theta} = u_0 \sin\theta \left[1 + \frac{1}{2}\left(\frac{a}{r}\right)^3\right] \tag{6.41}$$

Making use of the potential and stream lines, it is possible to perform a material balance on the solute in a differential element of a "stream tube" to obtain (see [11])

$$\frac{\partial c}{\partial \phi} = \frac{\partial}{\partial \phi}\left(D_L \frac{\partial c}{\partial \phi}\right) + \frac{\partial}{\partial \psi}\left(D_T \omega^2 \frac{\partial c}{\partial \psi}\right) \tag{6.42}$$

where ω is the distance to the flow axis, and D_L and D_T are the longitudinal and transverse dispersion coefficients, respectively.

The boundary conditions to be observed in the integration of Eq. 6.42 are: (1) the solute concentration is equal to the background concentration, c_0, far away from the sphere; (2) the solute concentration is equal to the equilibrium concentration, $c = c^*$, on the surface of the sphere and (3) the concentration field is symmetric about the flow axis.

For very low fluid velocities, dispersion is the direct result of molecular diffusion, with $D_T = D_L = D'_m$, and the numerical solution presented by Carvalho et al. [11] applies. Those authors suggest that their results are well approximated (with an error of less than 1%) by

$$k = \varepsilon \frac{D'_m}{d_1}\left[4 + \frac{4}{5}(\mathrm{Pe}')^{2/3} + \frac{4}{\pi}\mathrm{Pe}'\right]^{1/2} \tag{6.43}$$

where $\mathrm{Pe}' = u_0 d_1 / D'_m$ is the Peclet number for the soluble sphere.

Now, by substituting the average mass transfer coefficient, given by Eq. 6.43, into Eq. 6.37, the following expression is obtained for the approximate validity criterion of theory developed above:

$$\frac{1}{\mathrm{Pe}'}\left(1 + \frac{\pi}{\mathrm{Pe}'} + \frac{\pi}{5(\mathrm{Pe}')^{1/3}}\right) > \left(\frac{d_1}{(1-\varepsilon)L}\right)^2 \tag{6.44}$$

Finally, with Eq. 6.44 we could predict the volumetric flowrates that guarantee saturation in the outlet stream, $Q = \mathrm{Pe}'\pi D^2 \varepsilon D'_m/(4d_1)$. However, an important aspect to consider is the dependence of Q on the effective molecular diffusion coefficient, D'_m. Fortunately, values of D'_m increase with temperature, and the value of D'_m, at room temperature or lower, is a good estimate.

6.3.2 Experimental Set-Up

Experiments were performed on the dissolution of spheres of benzoic acid, 2-naphthol and salicylic acid (6.0 mm of internal diameter), buried in beds of sand

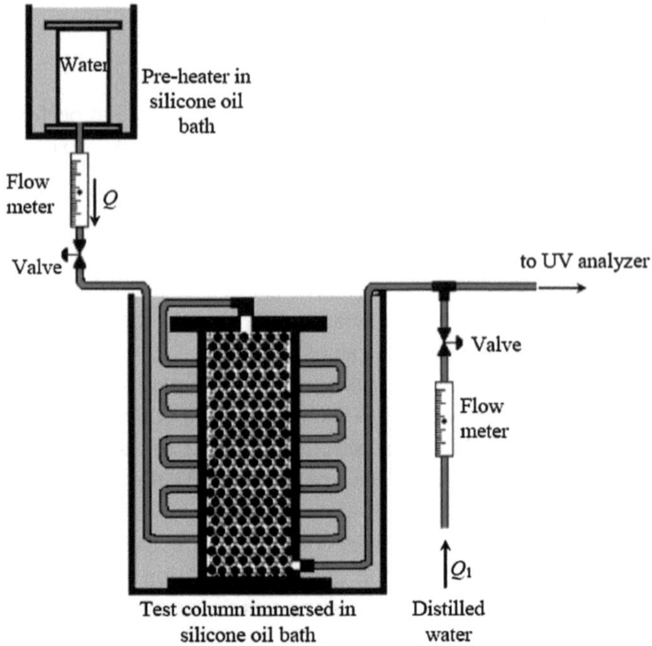

Fig. 6.10 Sketch of experimental set-up

(0.496 mm average particle diameter) through which water was steadily forced down, at temperatures in the range 293–373 K.

A stainless steel tube (21 mm i.d. and 200 mm long) was used to hold the bed of soluble solid spheres in an upright position while a metered stream of distilled water was fed to the top of the column, as sketched in Fig. 6.10. Near the bottom of the stainless steel column, a perforated plate, covered with fine wire mesh, was used to support the bed.

The distilled water was initially dearated, under vacuum, to avoid liberation of gas bubbles in the rig, at high temperature. The test column was immersed in a silicone oil bath kept at the desired operating temperature by means of a thermosetting bath head (not represented in Fig. 6.10). The copper tubing feeding the distilled water to the column at a constant metered rate was partly immersed in a pre-heater and it had a significant length immersed in the same thermosetting bath as the test column; the copper tubing leaving the test column was immersed in a chillier to cool the outlet stream before reaching the UV analyser.

The water flowrate was then adjusted to the required value, Q, and the concentration of solute in the outlet stream was continuously monitored by means of a UV/VIS Spectrophotometer (set at 274 nm, for 2-naphtol, at 226 nm, for benzoic acid and 292 nm, for salicylic acid).

The solubility of the solutes studied in water was calculated from the steady state average concentration of solute, c_{out}, in the outlet stream (refrigerated to room temperature), as $c^* = (1 + Q_1/Q)c_{\text{out}}$, where Q and Q_1 were the measured volumetric flowrates.

The spheres of solutes studied were prepared from *p.a.* grade material, which was molten and then poured into moulds made of silicone rubber. Wherever any slight imperfections showed on the surface of the spheres, they were easily removed by rubbing with fine sand paper. Using callipers three measurements were made of the diameter of each sphere along three perpendicular directions.

6.4 Measurement of Tortuosity in Porous Media

The two major properties to describing porous media and the associated mass transfer phenomena are the permeability coefficient (flow phenomena) and the effective diffusion (mass transfer phenomena). However, both coefficients are functions of the characteristics of porous media, namely the media porosity and tortuosity (see [17, 18]).

The tortuosity of a packed bed is an important parameter that describes pore connectivity and fluid transport, however, it is difficult to determine experimentally. Normally, tortuosity is calculated through measured values of the porosity and the experimentally determined effective diffusion coefficient. However, tortuosity varied with the particle volume fraction and the particle size ratio in the mixture.

In recent years, nuclear magnetic resonance (NMR) has been used to determine both diffusion and tortuosity coefficients (see *e.g.* [22, 36]), with significant advantages, but it's a very expensive method.

The experiment proposed is simple and inexpensive, and it provides an accurate method for the measurement of the tortuosity in packed beds.

Tortuosity can be defined as the ratio of the distance actually travelled by a tracer through the pore space, L_e, to the straight-line distance between the two points, L, of the porous media with solute concentrations C_1 and C_2. When ignoring that the "real" length (or effective) of diffusion trajectory is sinuous the representative equation of the diffusion law it should be written as:

$$Q = D'_{\text{m}} \varepsilon A \frac{C_1 - C_2}{L} \tag{6.45}$$

where D'_{m} is the "effective" molecular diffusion coefficient. However, the applicability of Ficks first law to the "real" length of diffusion in a packed bed, L_e, result as

$$Q = D_{\text{m}} \varepsilon A \frac{C_1 - C_2}{L_e} \tag{6.46}$$

where D_{m} is the molecular diffusion coefficient. Now, by substituting Eq. 6.45 into 6.46 the following relation is obtained:

$$D'_{m} = \frac{D_{m}}{L_{e}/L} \tag{6.47}$$

The ratio L_e/L being the tortuosity factor and it is usually represented by τ [7, 18]. In this work an original method for measurement of tortuosity in packed beds is presented.

Consider a vertical column containing a packed bed of inert particles, and filled with liquid to some level above the top of the packed bed (liquid "pool"). If a concentrated salt solution is then poured into this liquid "pool", with uniform concentration C_{p0}, the tracer will gradually penetrate down the packed bed. The concentration of tracer in the liquid "pool" decrease gradually in the time, until a uniform concentration of tracer, equilibrium, is reached in the whole liquid.

Analysis of the process of salt diffusion may be made in analogy with the process of diffusion from a stirred solution of limited volume, described by Crank [13]. The concentration of the dissolver salt in the packed bed, C, will vary according to (Fick's Second Law)

$$\frac{\partial C}{\partial t} = D'_{m} \frac{\partial^2 C}{\partial x^2} \tag{6.48}$$

subject to the following initial and boundary conditions,

$$t = 0 \quad 0 \le x \le L \quad C = 0 \tag{6.49a}$$

$$t > 0 \quad x = 0 \quad \frac{\partial C}{\partial x} = 0 \tag{6.49b}$$

$$t > 0 \quad x = L \quad -D'_{m}\varepsilon \frac{\partial C}{\partial x} = \frac{V_p}{A} \frac{\partial C_p}{\partial t} \tag{6.49c}$$

where V_p is the volume of liquid in the "pool", above the bed of inerts and A is the cross section area of the packed bed.

The solution of Eq. 6.48 with the boundary conditions (6.49a–c) could be obtained through the use of Fourier transform. However, the solution given by Crank [13] is conveniently expressed as the ratio between the amount of tracer in the bed of inert particles at any time, M_t, and the corresponding amount of tracer after a sufficiently long time, equilibrium, M_∞ :

$$\frac{M_t}{M_\infty} = 1 - \sum_{n=0}^{\infty} \frac{2\alpha(1+\alpha)}{1+\alpha+\alpha^2 q_n^2} \exp\left(-\frac{D'_{m} q_n^2 t}{L^2}\right) \tag{6.50}$$

where α is the ratio between the volume in the "pool" and the volume in the packed bed ($\alpha = V_p/V$) and q_n are the non-zero positive roots of

$$\tan(q_n) = -\alpha q_n \tag{6.51}$$

As smaller is the value of $D'_{\mathrm{m}}t/L^2$ more terms in the series, in Eq. 6.50, are needed for a given accuracy. When more than three or four terms are need it is better to use an alternative form of solution. For most values of α, the simplest expression is

$$\frac{\mathrm{M_t}}{\mathrm{M_\infty}} = (1+\alpha)\left[1 - \exp\left(\frac{D'_{\mathrm{m}}t}{L^2\alpha^2}\right) \times \operatorname{erfc}\left(\frac{D'_{\mathrm{m}}t}{L^2\alpha^2}\right)^{0.5}\right] \tag{6.52}$$

The amount of tracer in the packed bed, at any time t, could be expressed by the following equation,

$$\mathrm{M_t} = V_{\mathrm{p}}\left(C_{\mathrm{p0}} - C_{\mathrm{p}}(t)\right) \tag{6.53}$$

where C_{p0} is the initial tracer concentration in the "pool"; and the equilibrium amount of tracer in the packed bed of inerts is given by

$$\mathrm{M_\infty} = V(C_\infty - C_0) = V_{\mathrm{p}}\left(C_{\mathrm{p0}} - C_{\mathrm{p\infty}}\right) \tag{6.54}$$

where C_0 is the initial concentration of tracer in the bed of inerts ($C_0 = 0$) and C_∞ and $C_{\mathrm{p\infty}}$ are the equilibrium salt (tracer) concentration ($t = \infty$), in the packed bed and in the liquid "pool", respectively ($C_\infty = C_{\mathrm{p\infty}}$). So, the equilibrium tracer concentration in packed bed could be determined as,

$$C_\infty = \frac{C_0 V + C_{\mathrm{p0}} V_{\mathrm{p}}}{V_{\mathrm{p}} + V} \approx \frac{C_{\mathrm{p0}} V_{\mathrm{p}}}{V_{\mathrm{p}} + V} \tag{6.55}$$

and, for experimental data, the expression of the ratio between the amount of tracer at any time t and the amount of tracer in equilibrium is given by:

$$\frac{\mathrm{M_t}}{\mathrm{M_\infty}} = \frac{(V_{\mathrm{p}} + V)\left(C_{\mathrm{p0}} - C_{\mathrm{p}}\right)}{V\left(C_{\mathrm{p0}} - C_0\right)} \approx \frac{(V_{\mathrm{p}} + V)\left(C_{\mathrm{p0}} - C_{\mathrm{p}}\right)}{V C_{\mathrm{p0}}} \tag{6.56}$$

In each experiment the value of α is constant and the values of C_{p} was measured at different times in order to determine the corresponding values of $\mathrm{M_t}/\mathrm{M_\infty}$, from the appropriate representation of Eq. 6.52 or 6.54; and the value of $D'_{\mathrm{m}}t/L^2$ corresponding to each value of $\mathrm{M_t}/\mathrm{M_\infty}$ could be determined. For each experimental data point a plot of $D'_{\mathrm{m}}t/L^2$ *vs.* t, was organized and the best straight line through the points and the origin determined. The plot gives a straight line with slope of $D'_{\mathrm{m}}t/L^2$ and intercept of origin, to yield the experimental value of effective molecular diffusion coefficient, D'_{m}. Finally, the value of tortuosity was obtained by $\tau = D_{\mathrm{m}}/D'_{\mathrm{m}}$.

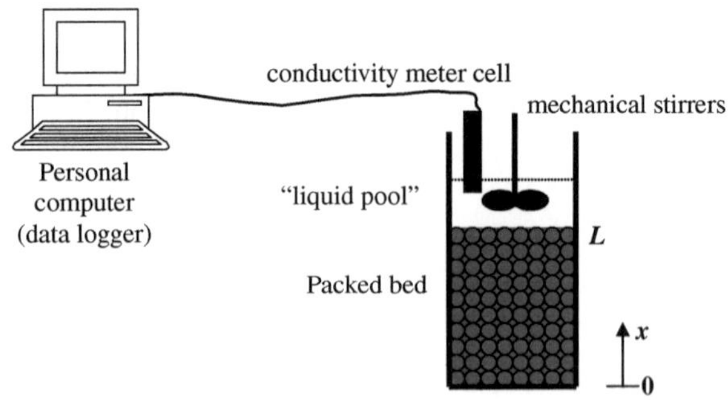

Fig. 6.11 Sketch of experimental set-up

6.4.1 Experimental Example

All measurements of tortuosity were performed in a transparent acrylic column, 0.10 m in diameter, was kept at a constant temperature using a thermostatically controlled water-bath (see Fig. 6.11).

The packed beds used in our experiments were beds of silica sand with average diameter of 0.496, 0.297, 0.219 and 0.110 mm. The silica sand was washed, dried and sieved in closely sized batches, for the experiments.

The dried silica sand was placed back inside the acrylic column at L length, and the packed bed was carefully compacted. Afterwards, distilled water, with volume V, was flowed in the "open space" of the packed bed; and a dilute salt solution flowed carefully on top of the packed bed, with volume V_p (liquid "pool"). A mechanical stirrer was used to homogenize the salt concentration in liquid "pool", with very slow rotation to avoid dispersion by forced convection.

The liquids used in the experiments were both distilled water and a dilute solution of NaCl in distilled water with a concentration of salt in the range of 0.05–2.0 M. The use of very dilute salt solutions and the care taken in equalizing the temperature of both liquids is needed to avoid dispersion by natural convection.

For each experiment the fraction of liquid volume in the packed bed, λ, was determined. The value of $\alpha = V_p/V$ used in the experiments is largely a matter of personal choice. The data in the present work refer to $\alpha \approx 1.5$.

To prevent evaporation the column are closed with a rubber cap. However, it is impossible to prevent some evaporation during the long execution time of the test.

The change of concentration in liquid "pool" was measured by monitoring decreases in the electrical conductance as a function of time. The output of this conductivity meter was connected to a micro-computer and care was taken to calibrate the conductivity cell.

6.5 Mass Transfer Around Active Solids

There are several situations of practical interest, both in nature and in man made processes, in which there is fluid flow through a bed of inert particles, packed around a solid mass that contacts with the moving fluid. Examples may be found in diverse fields, such as dilute catalyst fixed bed reactors, fluidised bed combustion, ore leaching and water contamination by buried waste. In such processes there is an interplay between diffusion, convection and dispersion and a detailed systematic study of the problem has been given by Coelho and Guedes de Carvalho [12] for transfer from buried flat surfaces and for transfer from buried spheres. Both these references present accurate solutions for certain limiting situations, namely those of low and high fluid velocity (more precisely, low and high Peclet numbers).

When the mass transfer process occur in a porous media with a fluid flowing around the soluble particle, and at low fluid velocities (as typically observed in underground flow), the assumption of thin boundary layer is not legitimate, and the theoretical analysis developed by Coelho and Guedes de Carvalho [12] is not applicable. Therefore, it is necessary to employ numerical methods for a correct analysis of the mass transfer process in more general situations.

Flow around a buried sphere is an important model situation in many processes and in a recent work Carvalho et al. [11] treated the problem numerically, so as to cover the entire range of values of Peclet and Schmidt numbers.

Flow along buried cylindrical surfaces and flat surfaces are also important model situations, and were investigated theoretically and numerically by Alves et al. [3], yielding results for a wide range of values of Peclet number, aspect ratio of soluble solid mass and Schmidt number.

Over the last years, our focus has been driven preferentially to the analysis of the mass transfer of particles with spherical geometry. This work is a contribution for the study of mass transfer of soluble particles with different geometries buried in inert particles with smaller diameter. Additionally, a simple approximate method is presented to obtain concentration contours plots for solute distribution around and downstream of the buried surfaces (with different geometries).

The following sections present a detailed description of the mass transfer and dispersion process around a soluble solid particle with different shapes (sphere, cylinder or a plane surface aligned with the flow, cylinder in cross-flow, prolate spheroid and a oblate spheroid) buried in a packed bed of smaller inert particles with uniform voidage, with a moving fluid with constant interstitial velocity.

6.5.1 Mass Transfer From a Soluble Flat Slab

Figure 6.12 sketches a section through a packed bed along which liquid is flowing, close to a flat wall, part of which ($0 < z < L$) is soluble. Liquid flow is assumed

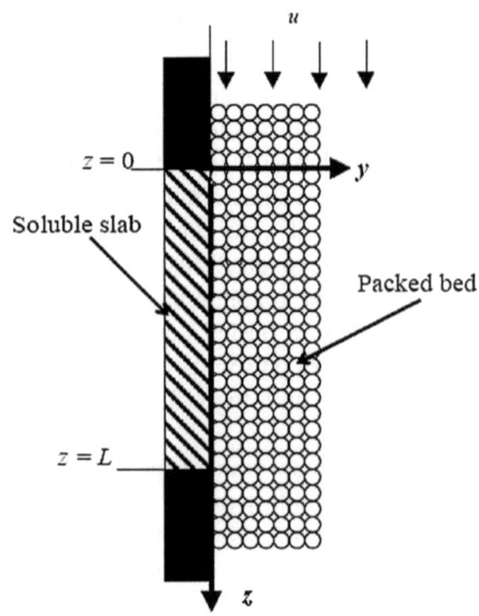

Fig. 6.12 Flow through packed bed near soluble flat surface

steady, with uniform average interstitial velocity u. If the concentration of solute in the liquid fed to the bed is c_0 and the solubility of the solid wall is c^*, a mass transfer boundary layer will develop, across which the solute concentration drops from $c = c^*$, at $y = 0$, to $c \to c_0$ for large y.

If we restrict the analysis to those situations for which the mass transfer boundary layer extends over several particle diameters and if a small control volume is considered, inside this boundary layer, with side lengths δz, δy and unity, a steady state material balance on the solute leads to

$$u\frac{\partial c}{\partial z} = D'_{\mathrm{m}}\frac{\partial^2 c}{\partial y^2} + D'_{\mathrm{m}}\frac{\partial^2 c}{\partial z^2} \tag{6.57}$$

To integrate Eq. 6.57 it is convenient to define the following dimensionless variables:

$$C = \frac{c - c_0}{c^* - c_0} \tag{6.58a}$$

$$Y = \frac{y}{L} \tag{6.58b}$$

$$Z = \frac{z}{L} \tag{6.58c}$$

$$\mathrm{Pe}'_{\mathrm{sf}} = \frac{uL}{D'_{\mathrm{m}}} \tag{6.58d}$$

where $\mathrm{Pe}'_{\mathrm{sf}}$ represents the Peclet number (based on the length, L, of the soluble slab), and D'_{m} is the effective molecular diffusion coefficient, defined as the ratio between the molecular diffusion coefficient and the tortuosity, τ, of the packed bed ($D'_{\mathrm{m}} = D_{\mathrm{m}}/\tau$).

In terms of dimensionless variables, Eq. 6.57 becomes

$$\mathrm{pe}'_{\mathrm{sf}} \frac{\partial C}{\partial Z} = \frac{\partial^2 C}{\partial Z^2} + \frac{\partial^2 C}{\partial Y^2} \tag{6.59}$$

and the appropriate boundary conditions are

$$\mathrm{C} \to 0 \qquad \mathrm{Z} \to -\infty \quad \forall \mathrm{Y} \tag{6.60a}$$

$$\mathrm{C} = 1 \qquad 0 \leq \mathrm{Z} \leq 1 \qquad Y = 0 \tag{6.60b}$$

$$\frac{\partial \mathrm{C}}{\partial \mathrm{Y}} = 0 \qquad \mathrm{Z} < 0 \vee \mathrm{Z} > 1 \qquad Y = 0 \tag{6.60c}$$

$$\mathrm{C} \to 0 \qquad \forall \mathrm{Z} \qquad \mathrm{Y} \to +\infty \tag{6.60d}$$

$$\mathrm{C} \to 0 \qquad \mathrm{Z} \to +\infty \qquad \mathrm{Y} \geq 0 \tag{6.60e}$$

Equation 6.59 is to be solved numerically, subjected to the boundary conditions (6.60a–6.60e), over the ranges of $\mathrm{Pe}'_{\mathrm{sf}}$ of practical interest.

6.5.1.1 Discretisation

Equation 6.59 was solved numerically, using a finite-difference method in a non-uniform grid similar to that adopted by Carvalho et al. [11]. A second-order central differencing scheme was adopted for the discretisation of the diffusive terms on the right hand side of Eq. 6.59, and the convective term, on the left hand side of Eq. 6.59, was discretised using the CUBISTA high-resolution scheme of Alves et al. [3], which preserves boundedness, even for highly advective flows.

The discretised equation resulting from the finite-difference approximation of Eq. 6.59 reads:

$$\mathrm{pe}'_{\mathrm{sf}} \frac{C_{i+1/2,j} - C_{i-1/2,j}}{(\Delta Z_i + \Delta Z_{i+1})/2} = \frac{C_{i+1,j}(\Delta Z_i) - C_{i,j}(\Delta Z_i + \Delta Z_{i+1}) + C_{i-1,j}(\Delta Z_{i+1})}{\Delta Z_i\, \Delta Z_{i+1}(\Delta Z_i + \Delta Z_{i+1})/2} + \frac{C_{i,j+1}(\Delta Y_j) - C_{i,j}(\Delta Y_j + \Delta Y_{j+1}) + C_{i,j-1}(\Delta Y_{j+1})}{\Delta Y_j\, \Delta Y_{j+1}(\Delta Y_j + \Delta Y_{j+1})/2} \tag{6.61}$$

The $C_{i+1/2,j}$ and $C_{i-1/2,j}$ values are interpolated from the known grid node values using the CUBISTA high-resolution scheme (HRS), in order to ensure numerical stability and good precision. The normalised variable approach (NVA) of Leonard [34] was adopted for implementations of the HRS, in which a general differencing scheme up to third order of accuracy can be expressed as

$$C_{i+1/2,j} = f(C_{i-1,j}, C_{i,j}, C_{i+1,j}) \tag{6.62}$$

The NVA uses an appropriate upwind biased normalisation, and Eq. 6.62 can be rewritten in compact form as

$$\widehat{C}_{i+1/2,j} = f(\widehat{C}_{i,j}) \tag{6.63}$$

where

$$\widehat{C}_{k,j} = \frac{C_{k,j} - C_{i-1,j}}{C_{i+1,j} - C_{i-1,j}} \quad (\text{for } k = i-1,\ i,\ i+1/2,\ i+1) \tag{6.64}$$

since, by definition of Eq. 6.64, $\widehat{C}_{i-1,j} = 0$ and $\widehat{C}_{i+1,j} = 1$.

The CUBISTA scheme is represented in the context of the NVA by the following piecewise linear functions [3]:

$$\widehat{C}_{i+1/2,j} = \begin{cases} \frac{7}{4}\widehat{C}_{i,j} & 0 \le \widehat{C}_{i,j} < \frac{3}{8} \\ \frac{3}{4}\widehat{C}_{i,j} + \frac{3}{8} & \frac{3}{8} \le \widehat{C}_{i,j} \le \frac{3}{4} \\ \frac{1}{4}\widehat{C}_{i,j} + \frac{3}{4} & \frac{3}{4} < \widehat{C}_{i,j} \le 1 \\ \widehat{C}_{i,j} & \text{elsewhere} \end{cases} \tag{6.65}$$

The resulting system of equations was solved iteratively using the successive over-relaxation (SOR) method [20], and the implementation of the boundary conditions was carried out in the same way as described in our previous work [11]. For the situation under study an orthogonal mesh is adequate and care was taken to ensure proper refinement in the regions where high concentration gradients are expected. The computational domain was defined according to the flow conditions (typically for small $\mathrm{Pe}'_{\mathrm{sf}}$, longer meshes are needed) and during mesh refinement, the number of nodes along each direction was doubled, thus halving the mesh sizes along each direction. This systematic procedure allows direct use of Richardson's extrapolation technique in order to obtain very accurate results [20].

6.5.1.2 Numerical Results

The numerical solution of Eq. 6.59 gives the concentration field and from it, the rate of dissolution of the slab, n, is obtained integrating the diffusion/dispersion flux over the whole slab surface. This integral is evaluated numerically, for each

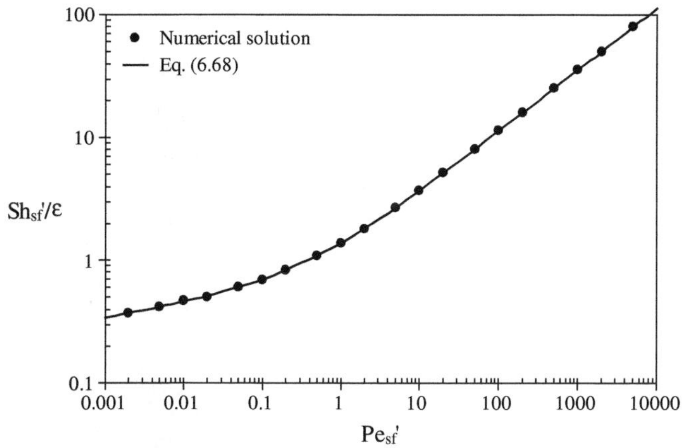

Fig. 6.13 Dependence of $\mathrm{Sh}'_{sf}/\varepsilon$ on Pe'_{sf} when $D_T = D_L = D'_m$ throughout, for slab flat

set of conditions, from the discretised concentration field that is obtained through the numerical solution of Eq. 6.59. It is convenient to express the rate of dissolution in terms of a Sherwood number, $\mathrm{Sh}'_{sf} = kL/D'_m$, where $k = n/[S(c^* - c_0)]$ is the mass transfer coefficient for the soluble slab and A is the exposed area of the soluble solid.

In the analysis of the results of the numerical computations we only consider the situation of low $\mathrm{Pe}'_p(= ud/D'_m)$ values, diffusional regime.

For low Pe'_p values, dispersion is the direct result of molecular diffusion and $D_T \cong D_L \cong D'_m$. Fig. 6.13 presents the results obtained numerically, for the situation of dispersion dominated by molecular diffusion, and two asymptotes can be observed:

$$\frac{\mathrm{Sh}'_{sf}}{\varepsilon} \to \left(\frac{2}{\pi}\right)^{1/2} \left(\mathrm{Pe}'_{sf}\right)^{1/8} \text{ for } \mathrm{Pe}'_{sf} \to 0 \tag{6.66}$$

and

$$\frac{\mathrm{Sh}'_{sf}}{\varepsilon} \to \left(\frac{4}{\pi}\right)^{1/2} \left(\mathrm{Pe}'_{sf}\right)^{1/2} \qquad \text{for } \mathrm{Pe}'_{sf} \to \infty \tag{6.67}$$

Taking the geometric mean of those two asymptotes

$$\frac{\mathrm{Sh}'_{sf}}{\varepsilon} = \left[\frac{2}{\pi}\mathrm{Pe}'^{\,1/4}_{sf} + \frac{4}{\pi}\mathrm{Pe}'_{sf}\right]^{1/2} \tag{6.68}$$

one observes that it does not deviate more than 3% from the numerical values.

Fig. 6.14 Flow through packed bed near soluble cylinder aligned with flow

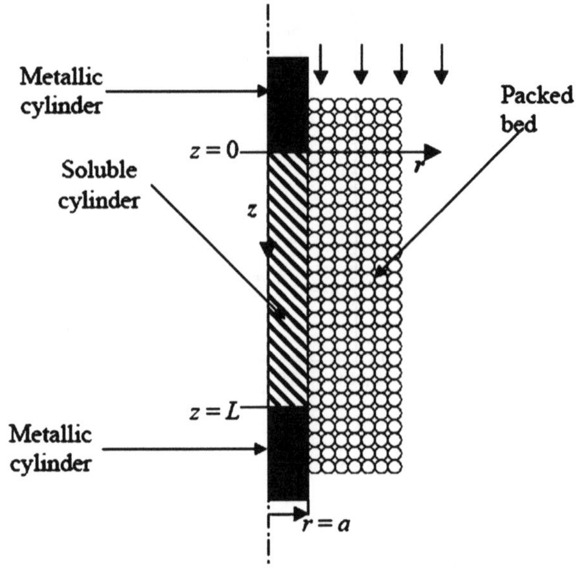

6.5.2 *Mass Transfer From a Soluble a Cylinder Aligned With Flow*

Figure 6.14 sketches a dissolving cylinder in a packed bed, aligned with the flow direction. In this case the mass transfer boundary layer forms around the cylinder and has axial symmetry.

Considering a small control-volume inside the boundary layer and performing a steady-state material balance on the solute, one obtains

$$\mathrm{Pe}_c' \frac{\partial C}{\partial Z} = \frac{\partial^2 C}{\partial Z^2} + 4\left(\frac{L}{d_1}\right)^2 \frac{1}{R}\frac{\partial}{\partial R}\left(R\frac{\partial C}{\partial R}\right) \tag{6.69}$$

where $R = 2r/d_c$, and d_c represents the diameter of the soluble cylinder. Equation 6.62 was solved numerically with boundary conditions

$$\mathrm{C} \to 0 \qquad \mathrm{Z} \to -\infty \qquad \mathrm{R} \geq 1 \tag{6.70a}$$

$$\mathrm{C} = 1 \qquad 0 \leq \mathrm{Z} \leq 1 \qquad \mathrm{R} = 1 \tag{6.70b}$$

$$\frac{\partial \mathrm{C}}{\partial \mathrm{R}} = 0 \qquad \mathrm{Z} < 0 \vee \mathrm{Z} > 1 \qquad \mathrm{R} = 1 \tag{670c}$$

$$\mathrm{C} \to 0 \qquad \forall \mathrm{Z} \qquad \mathrm{R} \to +\infty \tag{6.70d}$$

$$\mathrm{C} \to 0 \qquad \mathrm{Z} \to +\infty \qquad \mathrm{R} \geq 1 \tag{6.70e}$$

over the ranges of $\mathrm{Pe}_c' (= uL/D_m')$ and L/d_c of practical interest.

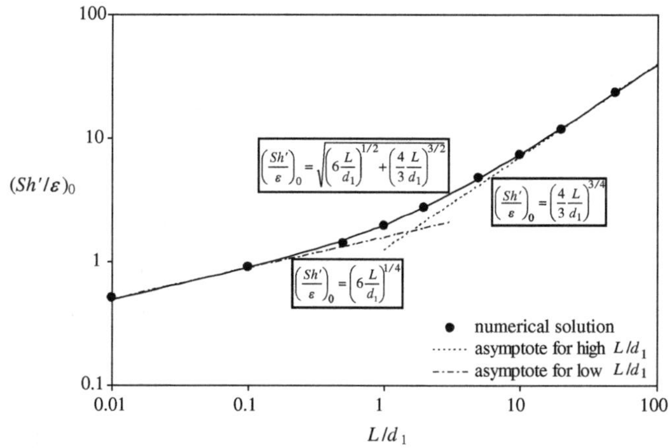

Fig. 6.15 Dependence of $(Sh'/\varepsilon)_0$ on L/d_1

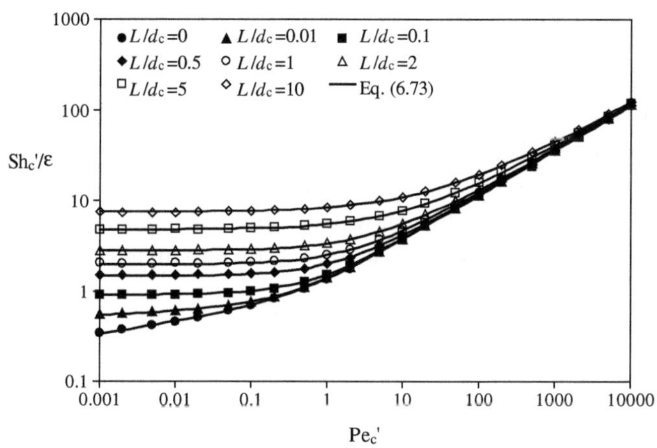

Fig. 6.16 Dependence of $\mathrm{Sh}_c'/\varepsilon$ on Pe_c' for different values of L/d_c

6.5.2.1 Numerical Results

For low Pe_p' values (say, $\mathrm{Pe}_p' < 0.1$), the values of $\mathrm{Sh}_c'/\varepsilon (= kL/D_m'\varepsilon)$ were obtained by integrating the flux along the soluble surface, in a similar way to that described for the flat surface. The first point to be mentioned is that the lower set of points in Fig. 6.16 (those for $L/d_c = 0$) were not obtained by solving Eq. 6.69 directly. Indeed, for $L/d_c \to 0$ (with finite L) the physical situation depicted in Fig. 6.15 degenerates into mass transfer from a flat surface. The plot for $L/d_c = 0$ reveals the two asymptotes obtained for the dissolution of a slab in a packed bed (see Eqs. 6.66 and 6.67. Equation 6.67 is the result obtained by Coelho and Guedes de

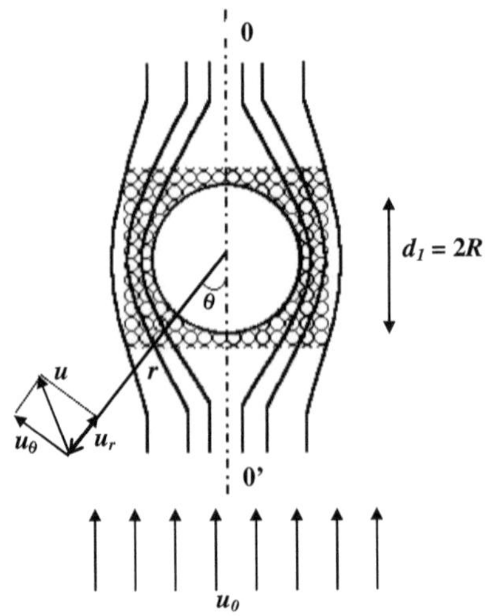

Fig. 6.17 Flow through packed bed near soluble sphere

Carvalho [12] who argued that it would be also the asymptote for $L/d_c > 0$. This is shown to be the case in Fig. 6.16 and it is not surprising, since for thin concentration boundary layer, the curvature of the cylinder is not relevant.

For each value of $L/d_c > 0$, the plot of $\mathrm{Sh}'_c/\varepsilon$ *vs.* Pe'_c has a horizontal asymptote, for $\mathrm{Pe}'_c \to 0$. This limiting value of $\mathrm{Sh}'_c/\varepsilon$ (for each value of L/d_c) may be conveniently represented by $\left(\mathrm{Sh}'_c/\varepsilon\right)_0$. to emphasize that it corresponds to the situation of pure molecular diffusion with no flow. The numerical results indicated a dependence of $\left(\mathrm{Sh}'_c/\varepsilon\right)_0$ on L/d_c and again, two asymptotes are revealed:

$$\left(\frac{\mathrm{Sh}'_c}{\varepsilon}\right)_0 \to \left(6\frac{L}{d_c}\right)^{1/4} \quad \text{for low } \frac{L}{d_c} \tag{6.71}$$

and

$$\left(\frac{\mathrm{Sh}'_c}{\varepsilon}\right)_0 \to \left(\frac{4}{3}\frac{L}{d_c}\right)^{3/4} \quad \text{for high } \frac{L}{d_c} \tag{6.72}$$

Having disclosed the above asymptotes, a general approximate expression for $\mathrm{Sh}'_c/\varepsilon$, that has the correct asymptotic behaviour, was found to be [4]

$$\frac{\mathrm{sh}'_c}{\varepsilon} = \left[\begin{array}{c} \frac{2}{\pi}\mathrm{pe}'^{\,1/4}_c + \frac{4}{\pi}\mathrm{pe}'_c + \left(6\frac{L}{d_c}\right)^{1/2} + \left(\frac{4}{3}\frac{L}{d_c}\right)^{3/2} \\ + \frac{5}{3}\mathrm{pe}'^{\,5/9}_c\left(\frac{L}{d_c}\right) - \mathrm{pe}'^{\,2/9}_c\left(2\frac{L}{d_c}\right)^{1/3} \end{array}\right]^{1/2} \tag{6.73}$$

This equation does not deviate more than 2% from the numerical values, over the entire range of values of $\mathrm{Pe}'_{\mathrm{c}}$ and $L/d_{\mathrm{c}}x$ represented in Fig. 6.16. The approximate conditions of validity of these equation are: $d_{\mathrm{c}}/d \geq 50$ and $L/d_{\mathrm{c}} \leq 100$.

6.5.3 Mass Transfer From a Soluble Sphere

Consider a soluble sphere of diameter $d_1\,(d_1 = 2a)$, buried in a bed of inert particles of diameter d (with $d \ll d_1$), packed uniformly (void fraction ε) around the sphere (see Fig. 6.17). The packed bed is assumed to be "infinite" in extent and a uniform interstitial velocity of liquid, u_0, is imposed, at a large distance from the sphere.

Darcy's law, $\boldsymbol{u} = -K\,\boldsymbol{grad}\,p$, is assumed to hold and if it is coupled with the continuity relation for an incompressible fluid, div $\boldsymbol{u} = 0$, Laplace's equation $\nabla^2\phi = 0$ is obtained for the flow potential, $\phi = Kp$, around the sphere. In terms of spherical coordinates (r, θ), the potential and stream functions are, respectively [14],

$$\frac{\partial c}{\partial \phi} = \frac{\partial}{\partial \phi}\left(D'_{\mathrm{m}}\frac{\partial c}{\partial \phi}\right) + \frac{\partial}{\partial \psi}\left(D'_{\mathrm{m}}\omega^2\frac{\partial c}{\partial \psi}\right) \tag{6.74}$$

where ω is the distance to the flow axis ($= r\sin\theta$).

To integrate Eq. 6.74, with the auxiliary Eqs. 6.75 and 6.76, it is convenient to define the following dimensionless variables:

$$\phi = -u_0\left[1 + \frac{1}{2}\left(\frac{a}{r}\right)^3\right] r\cos\theta \tag{6.75}$$

$$\psi = \frac{u_0}{2}\left[1 - \left(\frac{a}{r}\right)^3\right] r^2\sin^2\theta \tag{6.76}$$

and the velocity components are

$$u_r = \frac{\partial \phi}{\partial r} = -u_0\cos\theta\left[1 - \left(\frac{a}{r}\right)^3\right] \tag{6.77}$$

$$u_\theta = \frac{1}{r}\frac{\partial \phi}{\partial \theta} = u_0\sin\theta\left[1 + \frac{1}{2}\left(\frac{a}{r}\right)^3\right] \tag{6.78}$$

The analysis of mass transfer is based on a steady state material balance on the solute crossing the borders of an elementary volume, limited by the potential surfaces ϕ and $\phi + \delta\phi$, and the stream surfaces ψ and $\psi + \delta\psi$. The resulting equation is [4],

$$C = \frac{c - c_0}{c^* - c_0} \tag{6.79a}$$

$$\Re = \frac{r}{a} \tag{6.79b}$$

$$U = \frac{u}{u_0} = \frac{\left(u_r^2 + u_\theta^2\right)^{1/2}}{u_0} \tag{6.79c}$$

$$\Phi = \frac{4}{3}\frac{\phi}{u_0 d_1} \tag{6.79d}$$

$$\Psi = \frac{\psi}{u_0 d_1^2} \tag{6.79e}$$

Equation 6.78 may be re-arranged to

$$\frac{\partial C}{\partial \Phi} = \frac{\partial}{\partial \Phi}\left(\frac{4}{3\mathrm{Pe}_s'}\frac{\partial C}{\partial \Phi}\right) + \frac{\partial}{\partial \psi}\left(\frac{3}{16}\frac{\Re^2 \sin^2\theta}{\mathrm{Pe}_s'}\frac{\partial C}{\partial \psi}\right) \tag{6.80}$$

The boundary conditions to be observed in the integration of Eq. 6.74 are: (1) the solute concentration is equal to the background concentration, c_0, far away from the sphere; (2) the solute concentration is equal to the equilibrium concentration, $c = c^*$, on the surface of the sphere and (3) the concentration field is symmetric about the flow axis:

$$\Phi \to -\infty,\ \Psi \geq 0 \quad C \to 0 \tag{6.81a}$$

$$\Phi \to +\infty,\ \Psi \geq 0 \quad C \to 0 \tag{6.81b}$$

$$\Psi = 0 \begin{cases} -1 \leq \phi \leq 1 & C = 1 \\ |\phi| > 1 & \dfrac{\partial C}{\partial \Psi} = 0 \end{cases} \tag{6.81c, d}$$

$$\Psi \to +\infty,\ \text{all } \Phi \quad C \to 0 \tag{6.81e}$$

Equation 6.80 was solved numerically, subjected to the boundary conditions (6.81a–e), over the ranges of $\mathrm{Pe}_s' (= u_0 d_1 / D_m')$ of practical interest.

6.5.3.1 Numerical Results

For low values of Pe_p', dispersion is the direct result of molecular diffusion and the numerical solution obtained by Guedes de Carvalho and Alves [23] applies. Those authors showed that the numerical results are well described (with an error of less than 1% in $\mathrm{Sh}_s' = k d_1 / D_m'$) by (see Fig. 6.18)

$$\frac{\mathrm{Sh}_s'}{\varepsilon} = \left[4 + \frac{4}{5}\left(\mathrm{Pe}_s'\right)^{2/3} + \frac{4}{\pi}\mathrm{pe}_s'\right]^{1/2} \tag{6.82}$$

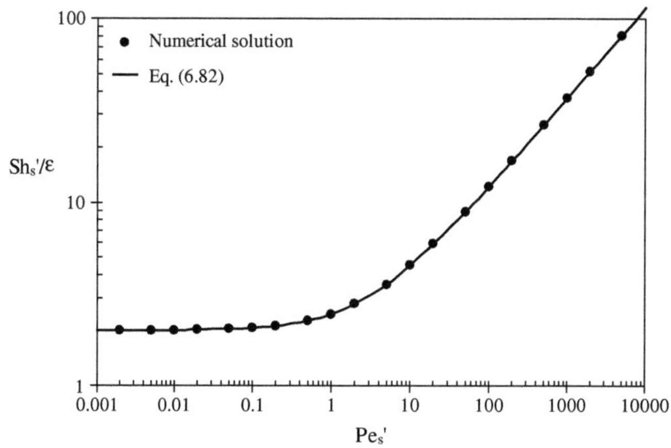

Fig. 6.18 Dependence of $\mathrm{Sh}_s'/\varepsilon$ on Pe_s' for soluble sphere

Fig. 6.19 Flow through packed bed near soluble cylinder in cross flow

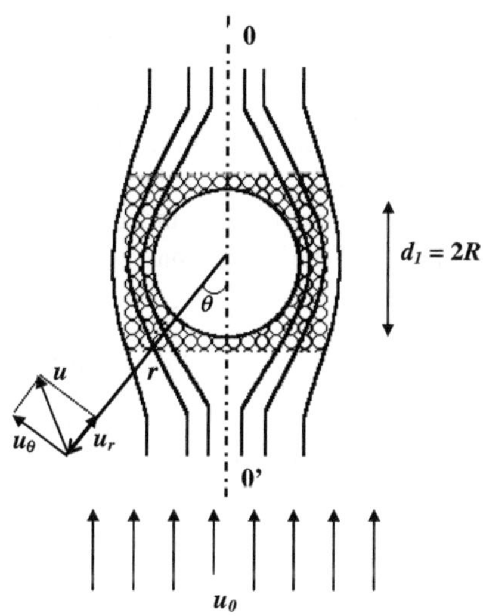

6.5.4 Mass Transfer From a Cylinder in Cross Flow

Consider a slightly soluble cylinder, of diameter d_1, buried in a packed bed of inert spherical particles of diameter d $(d \ll d_1)$ and exposed to fluid flow perpendicular to its axis, with uniform interstitial velocity, u_0, at a large distance from the cylinder (see Fig. 6.19).

Assuming Darcy's law, Laplace's equation is obtained for the flow potential around the cylinder. For a sufficiently long cylinder (assumed to be of "infinite" length) the flow field is bi-dimensional and in polar coordinates, the potential and stream functions are expressed as

$$\phi = -u_0\left[1 + \left(\frac{a}{r}\right)^2\right] r\cos\theta \tag{6.83}$$

$$\psi = u_0\left[1 - \left(\frac{a}{r}\right)^2\right] r\sin\theta \tag{6.84}$$

and the velocity components are given by

$$u_r = \frac{\partial\phi}{\partial r} = -u_0\cos\theta\left[1 - \left(\frac{a}{r}\right)^2\right] \tag{6.85}$$

$$u_\theta = \frac{1}{r}\frac{\partial\phi}{\partial\theta} = u_0\sin\theta\left[1 + \left(\frac{a}{r}\right)^2\right] \tag{6.86}$$

To formulate the mass transfer problem we take the concentration of the diffusing species to be c^* on the surface of the cylinder and c_0 at a large distance from it, in the approaching stream. The resulting concentration field will have axial symmetry and the differential equation for mass transfer may be derived from a mass balance on the elementary volume.

The analysis of mass transfer is based on a steady state material balance on the solute crossing the borders of an elementary volume, limited by the potential surfaces ϕ and $\phi+\delta\phi$, and the stream surfaces ψ and $\psi + \delta\psi$. The material mass balance on the solute may be expressed as

$$\frac{\partial c}{\partial\phi} = \frac{\partial}{\partial\phi}\left(D'_{\mathrm{m}}\frac{\partial c}{\partial\phi}\right) + \frac{\partial}{\partial\psi}\left(D'_{\mathrm{m}}\frac{\partial c}{\partial\psi}\right) \tag{6.87}$$

The assumption of steady state is acceptable for a solid that is taken to be slightly soluble and the use of the dispersion coefficients makes sense if the boundary layer extends over several particle diameters.

To integrate Eq. 6.87, with the auxiliary Eqs. 6.83 and 6.84, it is convenient to define the dimensionless variables:

$$C = \frac{c - c_0}{c^* - c_0} \tag{6.88a}$$

$$\Re = \frac{r}{R} \tag{6.88b}$$

$$U = \frac{u}{u_0} = \frac{\left(u_r^2 + u_\theta^2\right)^{1/2}}{u_0} \tag{6.88c}$$

$$\Phi = \frac{\phi}{u_0 d_{\mathrm{c}}} \tag{6.88d}$$

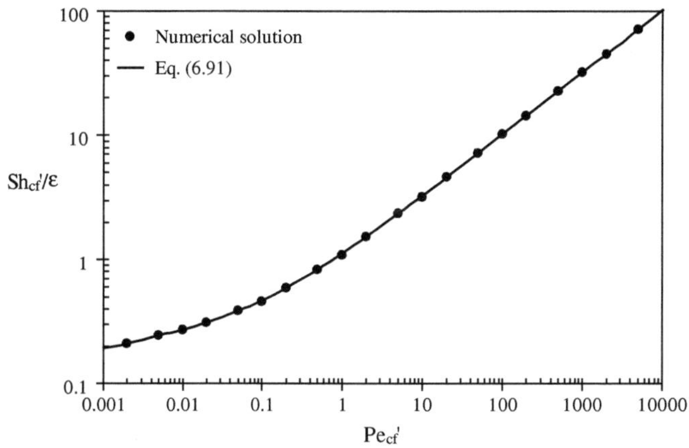

Fig. 6.20 Dependence of $\mathrm{Sh}'_{\mathrm{cf}}/\varepsilon$ on $\mathrm{Pe}'_{\mathrm{cf}}$, for a soluble cylinder in cross flow

$$\Psi = \frac{\psi}{u_0 d_c} \tag{6.88e}$$

$$\mathrm{Pe}'_{\mathrm{cf}} = \frac{u_0 d_c}{D'_m} \tag{6.88f}$$

where $\mathrm{Pe}'_{\mathrm{cf}}$ represents the Peclet number based on the diameter of the soluble cylinder, d_c. Equation 6.87 may then be rewritten as

$$\frac{\partial C}{\partial \Phi} = \frac{\partial}{\partial \Phi}\left(\frac{1}{\mathrm{Pe}'_{\mathrm{cf}}}\frac{\partial C}{\partial \Phi}\right) + \frac{\partial}{\partial \psi}\left(\frac{1}{\mathrm{Pe}'_{\mathrm{cf}}}\frac{\partial C}{\partial \psi}\right) \tag{6.89}$$

The boundary conditions to be observed in the integration of Eq. 6.89 are: (1) the solute concentration is equal to the background concentration, c_0, far from the cylinder; (2) the solute concentration is equal to the equilibrium concentration, $c = c^*$, on the surface of the cylinder and (3) the concentration field is symmetric about $\psi = 0$.

$$\Phi \to -\infty,\ \Psi \geq 0 \quad C \to 0 \tag{6.90a}$$

$$\Phi \to +\infty,\ \Psi \geq 0 \quad C \to 0 \tag{6.90b}$$

$$\Psi = 0 \begin{cases} -1 \leq \Phi \leq 1 & C = 1 \\ |\Phi| > 1 & \dfrac{\partial C}{\partial \Psi} = 0 \end{cases} \tag{6.90c, d}$$

$$\Psi \to +\infty,\ \text{all } \Phi \quad C \to 0 \tag{6.90e}$$

Equation 6.89 has to be solved numerically and the method developed and described in detail by Alves et al. [3] was adapted, to obtain the solute concentration field around the dissolving cylinder.

Fig. 6.21 The prolate spheroidal coordinates system

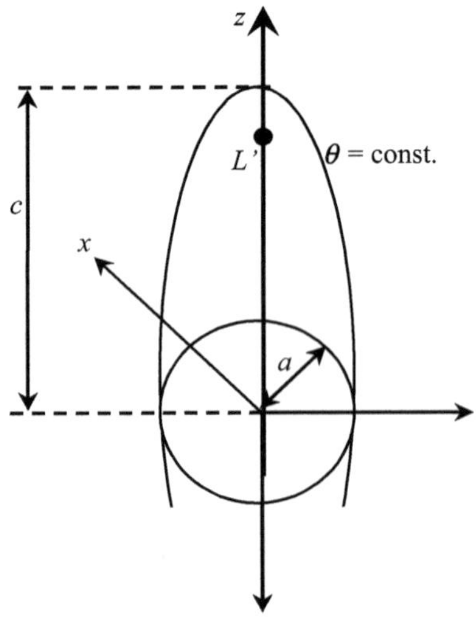

6.5.4.1 Numerical Results

For very low Pe'_p, dispersion is the direct result of molecular diffusion, with $D_T \cong D_L \cong D'_m$, and Sherwood number is only function of Pe′. Numerical solutions were worked out for the range of Pe′ between 10^{-3} and 10^4. The numerical values of Sh'/ε obtained are shown as dots in the plot of Fig. 6.20.

An improved approximation was found to describe the values of Sh'_{cf}/ε obtained numerically, Eq. 6.91, which is represented as a full line in the same figure:

$$\frac{Sh'_{cf}}{\varepsilon} = \left[\frac{2}{\pi^2} Pe'^{\,1/4}_{cf} + \frac{32}{\pi^3} Pe'_{cf}\right]^{1/2} \qquad (6.91)$$

and the values of Sh'_{cf}/ε calculated from this equation differ at most by 2% from the corresponding numerical solution obtained in the present work.

6.5.5 Mass Transfer From a Prolate Spheroid

In many practical situations it is often required to consider operations in which there are physico-chemical interactions between a solid particle and the fluid flowing around it. In the treatment of these operations it is common practice to assume the soluble particle to be spherical, because the treatment of irregular shapes could only be done by numerical methods.

The surface of the prolate spheroid ($\theta = \theta_0$) is described by (see Fig. 6.21)

$$\frac{x^2+y^2}{a^2}+\frac{z^2}{c^2}=1 \tag{6.92}$$

Since $r^2 = x^2 + y^2$, Eq. 6.92 can be written as

$$z = c\sqrt{1-(r/a)^2} \tag{6.93}$$

The surface area S and volume V of a prolate spheroid are given by

$$S = 2\pi a^2\left(1+\frac{c/a}{\sqrt{1-a^2/c^2}}\sin^{-1}\left(\sqrt{1-a^2/c^2}\right)\right) \tag{6.94}$$

$$V = \frac{4}{3}\pi a^2 c \tag{6.95}$$

where $e = \sqrt{1-a^2/c^2}$ is the ellipticity, and where $e = 0$ corresponds to a sphere. The dimensional Cartesian coordinates (x, y, z) are related to the prolate spheroidal ones (θ, η, β) through the equations [37]

$$x = L' \sinh\theta \sin\eta \cos\beta \tag{6.96a}$$

$$y = L' \sinh\theta \sin\eta \sin\beta \tag{6.96b}$$

$$z = L' \cosh\theta \cos\eta \tag{6.96c}$$

where L' is the focal distance $\left(L' = \sqrt{c^2-a^2}\right)$ and the coordinates range are: $0 \le \theta < \infty$, $0 \le \eta \le \pi$ and $0 \le \beta \le 2\pi$.

6.5.5.1 Analytical Results

For high Peclet numbers, with $D_T \cong D_L \cong D'_m$, our the theory is based on the assumption that the inert particles in the bed are packed with uniform voidage, ε, and that the flow may be approximated everywhere by Darcy's law, $\boldsymbol{u} = -K\,\boldsymbol{grad}\,p$. Furthermore, if the fluid is treated as incompressible, mass conservation leads to div $\boldsymbol{u} = 0$, Laplace's equation is obtained $\nabla^2\phi = 0$.

Darcy's law is strictly valid only for laminar flow through the packing, but according to Bear [7] it is still a good approximation for values of the Reynolds number (based on superficial velocity) up to ~ 10, which for beds with $\varepsilon \sim 0.4$ is equivalent to Re ~ 25, the upper limit for the validity of this analysis.

When a solid prolate spheroid is immersed in a packed bed of significantly smaller particles, through which fluid flows with uniform interstitial velocity u_0, far from the spheroid, the solution of Laplace's equation and the corresponding stream function, in terms of spheroidal coordinates (θ, η, β), are [2]

$$\phi = -u_0 L' \cos\eta \left[\cosh\theta - \frac{\cosh\theta \coth^{-1}(\cosh\theta) - 1}{\coth^{-1}(\cosh\theta_0) - \dfrac{\cosh\theta_0}{\sinh^2\theta_0}} \right] \tag{6.97}$$

$$\psi = u_0 \frac{L^2}{4} \sinh\theta \left[\sinh\theta - \cos(2\eta)\sinh\theta - (1-\cos(2\eta)) \frac{\sinh\theta \coth^{-1}(\cosh\theta) - \coth\theta}{\coth^{-1}(\cosh\theta_0) - \dfrac{\cosh\theta_0}{\sinh^2\theta_0}} \right] \tag{6.98}$$

The stream and potential functions are related to the dimensionless velocity components (u_θ, u_η) by Batchelor, [5]

$$u_\theta = \frac{1}{L'\sqrt{\sinh^2\theta + \sin^2\eta}} \frac{\partial\phi}{\partial\theta} = \frac{-1}{L'^2\sqrt{\sinh^2\theta + \sin^2\eta}\sinh\theta\sin\eta} \frac{\partial\psi}{\partial\eta} \tag{6.99}$$

$$u_\eta = \frac{1}{L'\sqrt{\sinh^2\theta + \sin^2\eta}} \frac{\partial\phi}{\partial\eta} = \frac{1}{L'^2\sqrt{\sinh^2\theta + \sin^2\eta}\sinh\theta\sin\eta} \frac{\partial\psi}{\partial\theta} \tag{6.100}$$

resulting the following velocity components

$$u_\theta = \frac{-u_0\cos\eta}{\sqrt{\sinh^2\theta + \sin^2\eta}} \left[\sinh\theta - \frac{\sinh\theta \coth^{-1}(\cosh\theta) - \coth(\theta)}{\coth^{-1}(\cosh\theta_0) - \dfrac{\cosh\theta_0}{\sinh^2\theta_0}} \right] \tag{6.101}$$

$$u_\eta = \frac{u_0\sin\eta}{\sqrt{\sinh^2\theta + \sin^2\eta}} \left[\cosh\theta - \frac{\cosh\theta \coth^{-1}(\cosh\theta) - 1}{\coth^{-1}(\cosh\theta_0) - \dfrac{\cosh\theta_0}{\sinh^2\theta_0}} \right] \tag{6.102}$$

The tangential velocity at the surface of the prolate spheroid $(\theta = \theta_0)$ can be found by

$$u_{\eta 0} = \frac{1}{L'\sqrt{\sinh^2\theta + \sin^2\eta}} \left. \frac{\partial\phi}{\partial\eta} \right|_{\theta=\theta_0} \tag{6.103}$$

and the resulting expression is

$$u_{\eta 0} = \frac{u_0\sin\eta}{(1/e^2 - 1 + \sin^2\eta)^{0.5}\left[1/e - (1/e^2 - 1)\tanh^{-1}e\right]} \tag{6.104}$$

Figure 6.22 shows the adimensional tangential surface velocity, $u_{\eta 0}/u_0$, of a prolate spheroid as a function of η, for different values of the eccentricity, e. Note that, for the case of a sphere, $e \approx 0$, the well-known result of $u_{\eta 0} = 1.5u_0\sin\eta$ is

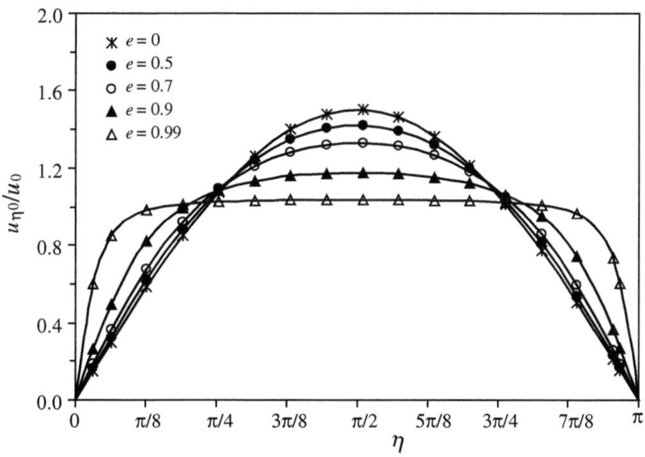

Fig. 6.22 The adimensional tangential surface velocity, $u_{\eta 0}/u_0$, as a function of η, for different values of the eccentricity, e

obtained, for potential flow over the surface of the sphere. On the other hand, for a slender prolate, $e \to 1$, as expected $u_{\eta 0}/u_0 \to 1$.

A convenient way of expressing the differential mass balance on the solute is to take a control volume along a stream tube, between two nearby potential surfaces. The resulting expression, for convection with molecular diffusion, is

$$\frac{\partial C}{\partial \phi} = \frac{\partial}{\partial \phi}\left(D'_{\mathrm{m}} \frac{\partial C}{\partial \phi}\right) + \frac{\partial}{\partial \psi}\left(D'_{\mathrm{m}} \omega^2 \frac{\partial C}{\partial \psi}\right) \tag{6.105}$$

For high values of the Peclet number the concentration boundary layer will be thin and the first term on the r.h.s. of Eq. 6.105 may be neglected. After some algebraic manipulation and a suitable change of variables, it is then possible to obtain

$$\frac{\partial \mathrm{C}}{\partial \xi} = \frac{\partial^2 \mathrm{C}}{\partial \psi^2} \tag{6.106}$$

where ξ is defined by

$$\xi = \int_0^{\eta} L'^3 \sinh^2 \theta_0 \sin^2 \eta' (\sinh^2 \theta_0 + \sin^2 \eta')^{0.5}\, u_\eta D'_{\mathrm{m}} d\eta' \tag{6.107}$$

with $\omega = L' \sinh \theta_0 \sin \eta$. The boundary conditions for Eq. 6.105, in our problem, are

$$C = C_0 \quad \xi = 0 \quad \psi > 0 \tag{6.108a}$$

$$C = C^* \quad \xi > 0 \quad \psi = 0 \tag{6.108b}$$

$$C \to C_0 \quad \xi > 0 \quad \psi \to \infty \tag{6.108c}$$

and the corresponding solution is

$$\frac{\mathrm{C} - \mathrm{C_0}}{\mathrm{C^*} - \mathrm{C_0}} = 1 - erf\left(\frac{\psi}{2\sqrt{\xi}}\right) \tag{6.109}$$

The value of ξ varies over the surface of the spheroid. Now, for potential flow u_η is given by Eq. 6.104 over the surface of the spheroid ($\theta = \theta_0$) and the integral in (6.107) is

$$\xi = u_0 D'_\mathrm{m} \frac{c^3 e^3 (1 - e^2)}{e - (1 - e^2)\tanh^{-1} e} \left(\frac{2}{3} - \cos\eta + \frac{1}{3}\cos^3\eta\right) \tag{6.110}$$

The flux of solute at any point on the surface of the spheroid is

$$\begin{aligned} N = -D'_\mathrm{m}\varepsilon\left(\frac{\partial C}{\partial b}\right) &= -\frac{D'_\mathrm{m}\varepsilon}{L'\sqrt{\sinh^2\theta_0 + \sin^2\eta}}\left(\frac{\partial C}{\partial \theta}\right)_{\theta=\theta 0} \\ &= -D'_\mathrm{m}\varepsilon\, u_\eta L' \sinh\theta_0 \sin\eta \left(\frac{\partial C}{\partial \psi}\right)_{\psi=0} \end{aligned} \tag{6.111}$$

and from (6.104) it may be shown that $(\partial C/\partial\psi)_{\psi=0} = \left(-1/\sqrt{(\pi\xi)}\right)(C^* - C_0)$. The rate of dissolution of the spheroid in the region $0 < \eta < \eta_1$ will then be

$$\begin{aligned} n(\eta_1) &= \int_0^{\eta 1} N 2\pi L'^2 \sqrt{\sinh^2\theta_0 + \sin^2\eta}\, \sinh\theta_0 \sin\eta \, \mathrm{d}\eta \\ &= \int_0^{\xi} 2\pi\varepsilon (C^* - C_0)\,(1/\sqrt{\pi z}) dz = 4\sqrt{\pi}\,\varepsilon\,\xi^{1/2}(C^* - C_0) \end{aligned} \tag{6.112}$$

with $\xi(\eta_1)$given by (6.110). In particular, the total rate of dissolution of the spheroid, n_T, may be obtained taking $\eta_1 = \pi$. By definition, the average mass transfer coefficient, k, is

$$k = n_\mathrm{T} / \left[S\,(C^* - C_0)\right] \tag{6.113}$$

the resulting expression for k (from (6.112) and (6.113)) is

$$k = \frac{4\varepsilon\sqrt{\pi}}{2\pi a^2\left(1 + \frac{\sqrt{1/(1-e^2)}}{e}\sin^{-1}(e)\right)} \left(\frac{4}{3}\frac{c^3(1-e^2)e^3 u_0 D'_\mathrm{m}}{e - (1-e^2)\tanh^{-1} e}\right)^{1/2} \tag{6.114}$$

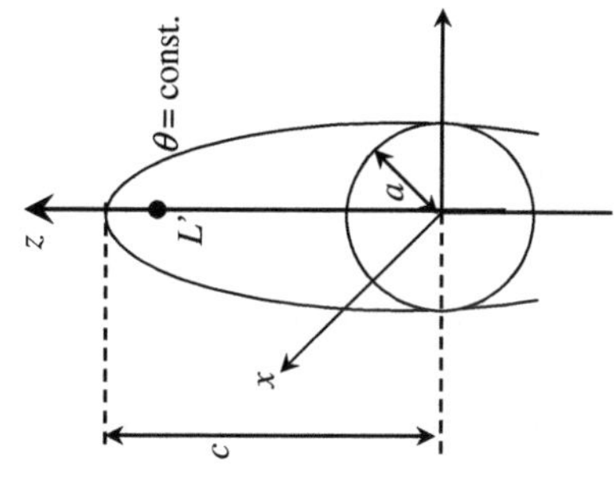

Fig. 6.23 The oblate spheroidal coordinates system

It is convenient to express the rate of dissolution in terms of the Sherwood number, $\mathrm{Sh}'_{\mathrm{ps}} = kd_{\mathrm{eq}}/D'_{\mathrm{m}}$, with $d_{\mathrm{eq}} = 2(a^2c)^{1/3}$, and the expression obtained, after some algebraic manipulation is

$$\frac{\mathrm{sh}'_{\mathrm{ps}}}{\varepsilon} = \sqrt{\frac{4}{\pi}\mathrm{Pe}'_{\mathrm{ps}}}\left(\frac{2}{3}\frac{e^3}{e-(1-e^2)\tanh^{-1}e}\right)^{1/2}\frac{2}{(1-e^2)^{1/3}+\frac{(1-e^2)^{-1/6}}{e}\sin^{-1}e} \tag{6.115}$$

where $\mathrm{Pe}'_{\mathrm{ps}} = u_0 d_{\mathrm{eq}}/D'_{\mathrm{m}}$ is the Peclet number. It is expected that Eq. 6.115 does not differ by more than 10% from the exact solution to obtain numerically.

For the special case of a sphere, $e \approx 0$, the result of $\mathrm{Sh}'_{\mathrm{s}}/\varepsilon = [4\ \mathrm{Pe}'/\pi]^{1/2}$ is obtained, which corresponds to the asymptotic behaviour for thin concentration layer (high values of Peclet number) when dispersion is constant and tend to D'_{m} over the surface of the sphere.

6.5.6 Mass Transfer From an Oblate Spheroid

The surface area S of an oblate spheroid is given by

$$S = 2\pi a^2\left(1+\frac{(c/a)^2}{2\sqrt{1-c^2/a^2}}\ln\left(\frac{1+\sqrt{1-c^2/a^2}}{1-\sqrt{1-c^2/a^2}}\right)\right) \tag{6.116}$$

where $e = \sqrt{1-c^2/a^2}$ is the ellipticity, and for $e = 0$ corresponds to a sphere. Figure 6.23 describes the oblate spheroidal coordinate system.

The dimensional Cartesian coordinates (x, y, z) are related to the oblate spheroidal ones (θ, η, β) through the equations

$$x = L'\cosh\theta\sin\eta\cos\beta \tag{6.117a}$$

$$y = L'\cosh\theta\sin\eta\sin\beta \tag{6.117b}$$

$$z = L' \sinh\theta \cos\eta \tag{6.117c}$$

where $L' = \sqrt{a^2 - c^2}$ is the focal distance and the coordinates are $0 \le \theta < \infty$, $0 \le \eta \le \pi$ and $0 \le \beta \le 2\pi$.

6.5.6.1 Analytical Results

For high values of Peclet numbers, the theory used is based on the assumption that the inert particles in the bed are packed with uniform voidage, ε, and that the fluid flow may be approximated everywhere by Darcy's law, $\mathbf{u} = -K\,\mathbf{grad}\,p$, as in the case of a prolate spheroid.

When a solid oblate spheroid is immersed in a packed bed of significantly smaller particles, through which fluid flows with uniform interstitial velocity u_0, far from the spheroid, the solution of Laplace's equation and the corresponding stream function, in terms of spheroidal coordinates (θ, η, β), are [2]

$$\phi = -u_0 L' \cos\eta \left[\sinh\theta - \frac{\sinh\theta\cot^{-1}(\sinh\theta) - 1}{\cot^{-1}(\sinh\theta_0) - \dfrac{\sinh\theta_0}{\cosh^2\theta_0}}\right] \tag{6.118}$$

$$\psi = u_0 \frac{L^2}{4}\cosh\theta \left[\begin{array}{l}\cosh\theta - \cos(2\eta)\cosh\theta \\ -(1-\cos(2\eta))\dfrac{\cosh\theta\cot^{-1}(\sinh\theta) - \tanh\theta}{\cot^{-1}(\sinh\theta_0) - \dfrac{\sinh\theta_0}{\cosh^2\theta_0}}\end{array}\right] \tag{6.119}$$

The stream and potential functions are related to the dimensionless velocity components and resulting the following velocity components

$$u_\theta = \frac{-u_0\cos\eta}{\sqrt{\cosh^2\theta - \sin^2\eta}}\left[\cosh\theta - \frac{\cosh\theta\cot^{-1}(\sinh\theta) - \tanh(\theta)}{\cot^{-1}(\sinh\theta_0) - \dfrac{\sinh\theta_0}{\cosh^2\theta_0}}\right] \tag{6.120}$$

$$u_\eta = \frac{u_0\sin\eta}{\sqrt{\cosh^2\theta - \sin^2\eta}}\left[\sinh\theta - \frac{\sinh\theta\cot^{-1}(\sinh\theta) - 1}{\cot^{-1}(\sinh\theta_0) - \dfrac{\sinh\theta_0}{\cosh^2\theta_0}}\right] \tag{6.121}$$

The tangential velocity at the surface of the oblate spheroid ($\theta = \theta_0$) can be found by

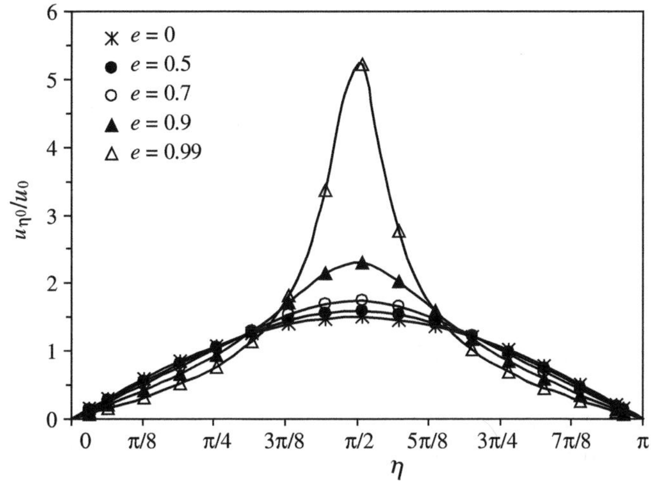

Fig. 6.24 The adimensional tangential surface velocity, $u_{\eta 0}/u_0$, as a function of η, for different values of the eccentricity, e

$$u_{\eta 0} = \frac{1}{L'\sqrt{\cosh^2\theta - \sin^2\eta}} \left.\frac{\partial \phi}{\partial \eta}\right|_{\theta=\theta_0}$$
$$= \frac{u_0 \sin\eta}{\left(\frac{1}{e^2} - \sin^2\eta\right)^{0.5}\left[-\sqrt{\frac{1-e^2}{e^2}} + \frac{1}{e^2}\cot^{-1}\left(\sqrt{\frac{1-e^2}{e^2}}\right)\right]} \tag{6.122}$$

Figure 6.24 shows the adimensional tangential surface velocity, $u_{\eta 0}/u_0$, of a prolate spheroid as a function of η, for different values of the eccentricity, e. Note that, for the case of a sphere, $e \approx 0$, the well-known result of $u_{\eta 0} = 1.5u_0 \sin\eta$ is obtained, for potential flow over the surface of the sphere. On the other hand, for a slender oblate, $e \to 1$, as expected $u_{\eta 0}/u_0 \to 1$.

A convenient way of expressing the differential mass balance on the solute is to take a control volume along a stream tube, between two nearby potential surfaces. The resulting expression, for convection with molecular diffusion, is given by Eq. 6.99, where ξ is defined by

$$\xi = \int_0^\eta L'^3 \cosh^2\theta_0 \sin^2\eta' (\cosh^2\theta_0 - \sin^2\eta')^{1/2} u_\eta D'_m \,\mathrm{d}\eta' \tag{6.123}$$

with $\omega = L' \cosh\theta_0 \sin\eta$. The boundary conditions for are given by Eqs. 6.108a–c and the corresponding solution is

$$\frac{C - C_0}{C^* - C_0} = 1 - erf\left(\frac{\psi}{2\sqrt{\xi}}\right) \tag{6.124}$$

The value of ξ varies over the surface of the spheroid. Now, for potential flow u_η is given by Eq. 6.115 over the surface of the spheroid ($\theta = \theta_0$) and the integral in (6.125) is

$$\xi = u_0 D'_{\mathrm{m}} \frac{e^3c^3/(1-e^2)}{-e + e^3 + \sqrt{1-e^2}\cot^{-1}(\sqrt{1/e^2-1})} \left(\frac{2}{3} - \cos\eta + \frac{1}{3}\cos^3\eta\right) \tag{6.125}$$

The flux of solute at any point on the surface of the spheroid is

$$N = -\frac{D'_{\mathrm{m}}\varepsilon}{L'\sqrt{\cosh^2\theta_0 - \sin^2\eta}} = -D'_{\mathrm{m}}\varepsilon\, u_\eta L' \cosh\theta_0 \sin\eta \left(\frac{\partial C}{\partial\psi}\right)_{\psi=0} \left(\frac{\partial C}{\partial\theta}\right)_{\theta=\theta_0} \tag{6.126}$$

and from (6.126) it may be shown that $(\partial C/\partial\psi)_{\psi=0} = \left(-1/\sqrt{(\pi\xi)}\right)(C^* - C_0)$. The rate of dissolution of the spheroid in the region $0<\eta<\eta_1$ will then be

$$n(\eta_1) = \int_0^{\eta 1} N 2\pi L'^2 \sqrt{\cosh^2\theta_0 - \sin^2\eta}\cosh\theta_0 \sin\eta\, \mathrm{d}\eta = 4\sqrt{\pi}\,\varepsilon\,\xi^{1/2}(C^* - C_0) \tag{6.127}$$

with $\xi(\eta_1)$given by (6.125). In particular, the total rate of dissolution of the spheroid, n_{T}, may be obtained taking $\eta_1 = \pi$. By definition, the average mass transfer coefficient, k, is given by Eq. 6.106 and the resulting expression for k is

$$k = \frac{4\varepsilon\sqrt{\pi}}{2\pi a^2\left(1 + \frac{1-e^2}{2e}\ln\left(\frac{1+e}{1-e}\right)\right)} \left(\frac{2}{3}\frac{e^3 u_0 D'_{\mathrm{m}} c^3/(1-e^2)}{-e + e^3 + \sqrt{1-e^2}\cot^{-1}(\sqrt{1/e^2-1})}\right)^{1/2} \tag{6.128}$$

It is convenient to express the rate of dissolution in terms of the Sherwood number, $\mathrm{Sh}'_{\mathrm{os}} = kd_{\mathrm{eq}}/D'_{\mathrm{m}}$, with $d_{\mathrm{eq}} = 2(a^2c)^{1/3}$, and the expression obtained is

$$\frac{\mathrm{Sh}'_{\mathrm{os}}}{\varepsilon} = \sqrt{\frac{4}{\pi}\mathrm{Pe}'_{\mathrm{os}}} \left(\frac{4}{3}\frac{e^3}{-e + e^3 + \sqrt{1-e^2}\cot^{-1}(\sqrt{1/e^2-1})}\right)^{1/2} \frac{(1-e^2)^{1/3}}{1 + \frac{1-e^2}{2e}\ln\left(\frac{1+e}{1-e}\right)} \tag{6.129}$$

where $\mathrm{Pe}'_{\mathrm{os}} = u_0 d_{\mathrm{eq}}/D'_{\mathrm{m}}$ is the Peclet number.

References

1. Akehata, T., Sato, K.: Flow distribution in packed beds. Chem. Eng. Jpn. **22**, 430–436 (1958)
2. Alassar, R.S., Badr, H.M.: Analytical solution of oscillating inviscid flow over oblate spheroids with spheres and flat disks as special cases. Ocean Eng. **24**, 217–225 (1997)
3. Alves, M.A., Oliveira, P.J., Pinho, F.T.: A convergent and universally bounded interpolation scheme for the treatment of advection. Int. J. Num. Meth. Fluids **41**, 47–75 (2003)
4. Alves, M.A., Delgado, J.M.P.Q., Guedes de Carvalho, J.R.F.: Mass transfer from cylinders and plane surfaces buried in packed beds in alignment with the flow direction. Chem. Eng. Sci. **61**, 1174–1183 (2006)
5. Batchelor, G.K.: An Introduction to Fluid Dynamics. Cambridge University Press, Cambridge (1997)
6. Baetsle, L.H.: Migration of radionuclides in porous media. In: Duhamed Pergamon, A.M.F. (ed.) Progress in Nuclear Energy, Series XII, pp. 707–730. Health Physics, Elmsford (1969)
7. Bear, J.: Dynamics of Fluids in Porous Media. Dover Publications, New York (1972)
8. Blackwell, R.J.: Laboratory studies of microscopic dispersion phenomena. Soc. Petrol. Eng. J. **225**, 1–8 (1962)
9. Bruinzeel, C, Reman, GH, van der Laan, ETh Axial dispersion. Proceedings of the 3th Congress of the European Federation of Chemical Engineering, Olympia (1962)
10. Carslaw, H.S., Jaeger, J.C.: Conduction of Heat in Solids, 2nd edn. Oxford University Press, Oxford (1959)
11. Carvalho, J.R.F.G., Delgado, J.M.P.Q., Alves, M.A.: Mass transfer between flowing fluid and sphere buried in packed bed of inerts. AIChE J. **50**, 65–74 (2004)
12. Coelho, M.A.N., Guedes de Carvalho, J.R.F.: Transverse dispersion in granular beds. Part I: Mass transfer from a wall and the dispersion coefficient in packed beds. Chem. Eng. Res. Des. **66**, 165–177 (1988)
13. Crank, J.: The Mathematics of Diffusion, 2nd edn. Oxford University Press, Oxford (1975)
14. Currie, I.G.: Fundamental Mechanics of Fluids. McGraw-Hill, New York (1993)
15. Danckwerts, P.V.: Continuous flow systems. Chem. Eng. Sci. **2**, 1–13 (1953)
16. Dorweiler, V.P., Fahien, R.W.: Mass transfer at low flow rates in a packed column. AIChE J. **5**, 139–144 (1959)
17. Dullien, F.A.L.: Single phase flow through porous media and pore structure. Chem. Eng. J. **10**, 1–34 (1975)
18. Dullien, F.A.L.: Porous Media: Fluid Transport and Pore Structure. Academic, San Diego (1979)
19. Fahien, R.W., Smith, J.M.: Mass transfer in packed beds. AIChE J. **1**, 28–37 (1955)
20. Ferziger, J.H., Peric, M.: Computational Methods for Fluid Dynamics. Springer, Berlin (1996)
21. Fuller, E.N., Schettler, P.D., Giddings, J.C.: A new method for prediction of binary gas-phase diffusion coefficients. Ind. Eng. Chem. **58**, 19–27 (1966)
22. Gibbs, S.J., Lightfoot, E.N., Root, T.W.: Protein diffusion in porous gel filtration chromatography media studied by pulsed field gradient NMR spectroscopy. J. Phys. Chem. **96**, 7458–7462 (1992)
23. Guedes de Carvalho, J.R.F., Alves, M.A.M.: Mass transfer and dispersion around active sphere buried in a packed bed. AIChE J. **45**, 2495–2502 (1999)
24. Guedes de Carvalho, J.R.F., Delgado, J.M.P.Q.: Lateral dispersion in liquid flow through packed beds at $Pe_m < 1400$. AIChE J. **46**, 1089–1095 (2000)
25. Gunn, D.J.: Mixing in packed and fluidised beds. Chem. Eng. **219**, CE153–CE172 (1968)
26. Gunn, D.J., Pryce, C.: Dispersion in packed beds. Trans. IChemE **47**, T341–T350 (1969)
27. Han, N.W., Bhakta, J., Carbonell, R.G.: Longitudinal and lateral dispersion in packed beds: effect of column length and particle size distribution. AIChE J. **31**, 277–288 (1985)
28. Harleman, D.R.F., Rumer, R.: Longitudinal and lateral dispersion in an isotropic porous medium. J. Fluid Mech. **16**, 1–12 (1963)

29. Harrison, D., Lane, M., Walne, D.J.: Axial dispersion of liquid on a column of spheres. Trans. IChemE **40**, 214–220 (1962)
30. Hiby, J.W., Schummer, P.: Zur messung der transversalen effektiven diffusion in durchstromten fullkorpersaulen. Chem. Eng. Sci. **13**, 69–74 (1960)
31. Higbie, S.: The rate of absorption of a pure gas into a still liquid during short periods of exposure. Trans. AIChE J. **31**, 365–389 (1935)
32. Hunt, B.: Dispersive sources in uniform groundwater flow. J. Hydraul. Div.–Proc. Am. Soc. Civil Eng. **104**, 75–85 (1978)
33. Klinkenberg, A., Krajenbrink, H.J., Lauwerier, H.A.: Diffusion in a fluid moving at uniform velocity in a tube. Ind. Eng. Chem. **45**, 1202–1208 (1953)
34. Leonard, B.P.: Simple high-accuracy resolution program for convective modelling of discontinuities. Int. J. Num. Meth. Fluids **8**, 1291–1318 (1988)
35. Marrero, T.R., Mason, E.A.: Gaseous diffusion coefficients. J. Phys. Chem. Ref. Data **1**, 3–25 (1972)
36. Mair, R.W., Hurlimann, M.D., Sen, P.N., Schwartz, L.M., Patz, S., Walsworth, R.L.: Tortuosity measurement and the effects of finite pulse widths on xenon gas diffusion NMR studies of porous media. Mag. Reson. Imag. **193**, 345–351 (2001)
37. Moon, P., Spencer, D.E.: Field theory handbook (2nd ed). Springer, Berlin (1971)
38. National Research Council: International critical tables–Volume 5. McGraw-Hill International Editions, New York (1929)
39. Ogata, A., Banks, R.B.: A solution of differential equation of longitudinal dispersion in porous media. US Geolog. Surv. Prof. Pap. **411-A**:7–12 (1961)
40. Perkins, T.K., Johnston, O.C.: A review of diffusion and dispersion in porous media. Soc. Petrol. Eng. J. **2**, 70–84 (1963)
41. Perry, R.H., Green, D.: Perry's chemical engineer's handbook (6th ed). McGraw-Hill International Editions, New York (1984)
42. Pinto, A.M.F.R., Guedes de Carvalho, J.R.F.: Transverse dispersion in granular beds. Part III: Mass transfer around particles dispersed in granular beds of inerts and the combustion of carbon particles in beds of sand. Trans. IChemE **68**, 503–509 (1980)
43. Plautz, D.A., Johnstone, H.F.: Heat and mass transfer in packed beds. AIChE J. **1**, 193–199 (1955)
44. Reid, R.C., Prausnitz, J.M., Poling, B.E.: The properties of gases and liquids (4th ed). McGraw-Hill International Editions, New York (1988)
45. Rifai, M.N.E., Kaufman, W.J., Todd, D.K.: Dispersion phenomena in laminar flow through porous media. University of California, Sanitary Engineering Report 3. Inst. Eng. Res. Series **90**, 1–157 (1956)
46. Robbins, G.A.: Methods for determining transverse dispersion coefficients of porous media in laboratory column experiment. Water Resour. Res. **25**, 1249–1258 (1989)
47. Roemer, G., Dranoff, J.S., Smith, J.M.: Diffusion in packed beds at low flow rates. Ind. Eng. Chem. Fund. **1**, 284–287 (1962)
48. Sherwood, T.K., Pigford, R.L., Wilke, C.R.: Mass Transfer. McGraw-Hill International Editions, Tokyo (1975)
49. Sinclair, R.J., Potter, O.E.: The dispersion of gas in flow through a bed of packed solids. Trans. IChemE **43**, T3–T9 (1965)
50. Stephenson, J.L., Stewart, W.E.: Optical measurements of porosity and fluid motion in packed beds. Chem. Eng. Sci. **41**, 2161–2170 (1986)
51. Sun, N.Z.: Mathematical modelling of groundwater pollution. Springer, New York (1996)
52. Taylor, G.: Dispersion of soluble matter in solvent flowing slowly through a tube. P. Roy Soc. A –Math. Phy. **219**, 186–203 (1953)
53. Uno, S.: Mass Transfer from naphthalene slabs. PhD Dissertation, Illinois Institute of Technology, Chicago (1958)
54. Van Genuchten, M.Th., Alves, W.J.: Analytical solutions of the one-dimensional convective-dispersive solute transport equation. Tech. Bull. U.S. Dep. Agric. **1661**, 149–165 (1982)

55. Vortmeyer, D., Schuster, J.: Evaluation of steady flow profiles in rectangular and circular packed beds by a variational method. Chem. Eng. Sci. **38**, 1691–1699 (1983)
56. Wakao, N., Funazkri, T.: Effect of fluid dispersion coefficients on particle-to-fluid mass transfer coefficients in packed beds. Chem. Eng. Sci. **33**, 1375–1384 (1978)

Chapter 7
Applications and Examples

7.1 Contaminant Plume Sizes Associated to Different Active Solids

There are several situations of practical interest, both in nature and in manmade processes, in which there is fluid flow through a bed of inert particles, packed around a solid mass that contacts with the moving fluid. Examples may be found in diverse fields, such as dilute catalyst fixed bed reactors, fluidized bed combustion, ore leaching and water contamination by buried waste. In such processes there is an interplay between diffusion, convection and dispersion and a detailed systematic study of the problem has been given by Coelho and Guedes de Carvalho [16] for transfer from buried flat surfaces and for transfer from buried spheres. Both these references present accurate solutions for certain limiting situations, namely those of low and high fluid velocity (more precisely, low and high Peclet numbers).

When the mass transfer process occurs in a porous media with a fluid flowing around the soluble particle, and at low fluid velocities (as typically observed in underground flow), the assumption of thin boundary layer is not legitimate, and the theoretical analysis developed by Coelho and Guedes de Carvalho [16] is not applicable. Therefore, it is necessary to employ numerical methods for a correct analysis of the mass transfer process in more general situations.

Flow around a buried sphere is an important model situation in many processes and in a recent work Guedes de Carvalho et al. [31] treated the problem numerically, so as to cover the entire range of values of Peclet and Schmidt numbers.

Flow along buried cylindrical surfaces and flat surfaces are also important model situations, and were investigated theoretically and numerically by Alves et al. [8], yielding results for a wide range of values of Peclet number, aspect ratio of soluble solid mass and Schmidt number.

Over the last years, our focus has been driven preferentially to the analysis of the mass transfer of particles with spherical geometry. This work is a contribution

F. A. Coutelieris and J. M. P. Q. Delgado, *Transport Processes in Porous Media*, Advanced Structured Materials 20, DOI: 10.1007/978-3-642-27910-2_7,

for the study of mass transfer of soluble particles with different geometries buried in inert particles with smaller diameter. Additionally, a simple approximate method is presented to obtain concentration contours plots for solute distribution around and downstream of the buried surfaces (with different geometries).

The following sections present a detailed description of the mass transfer and dispersion process around a soluble solid particle with different shapes (sphere, cylinder or a plane surface aligned with the flow, cylinder in cross-flow, prolate spheroid and a oblate spheroid) buried in a packed bed of smaller inert particles with uniform voidage, with a moving fluid with constant interstitial velocity.

7.1.1 Concentration Profiles From a Soluble Flat Slab

In many processes, both in industry and in the environment, there is a physical/chemical interaction between a lump of solid matter buried in a granular bed of small inert particles, and the fluid flowing around it, through the interstices in the bed. In such processes, a region of higher solute/reactant concentration develops in the immediate vicinity of the "active mass", with consequent solute migration away from it, by diffusion and convection in the moving fluid.

In groundwater flow applications the velocities are low and small values of Peclet number are therefore observed; solute dispersion is then determined by molecular diffusion.

The concentration distribution in the wake of a soluble solid particle immersed in a granular bed of inert particles, through which fluid flows with "uniform velocity", has been obtained, for solute transport by both advection and diffusion. General expressions, based in a theoretical analysis and using the expressions obtained for the mass transfer rate, n, are presented to estimate contaminant "plume" sizes downstream of the polluting source.

In this model situation, we only consider the steady state conditions when the plane surface releases solute to the flowing liquid at a constant rate, n. The value of n depends on the concentration gradients next to the surface of the plane surface and these depend on the process variables. According to Eq. 6.68, the following equation

$$n = \frac{D'_{\mathrm{m}}}{L} \varepsilon bL\left(c^{*} - c_{0}\right)\left[\frac{2}{\pi}\mathrm{Pe}'^{\,1/4}_{\mathrm{sf}} + \frac{4}{\pi}\mathrm{Pe}'_{\mathrm{sf}}\right]^{1/2} \tag{7.1}$$

gives n within 3% of the exact value, over the entire range of values of $\mathrm{Pe}'_{\mathrm{sf}}$, if $\mathrm{Pe}'_{\mathrm{p}} < 0.1$.

Imagine a situation of continuous injection of solute, at a rate n, at the point $(x = 0,\ y = 0)$, in the porous medium through which fluid flows with uniform interstitial velocity u_0. It is assumed that the volume of additional fluid injected into the main fluid, with the solute, is negligible.

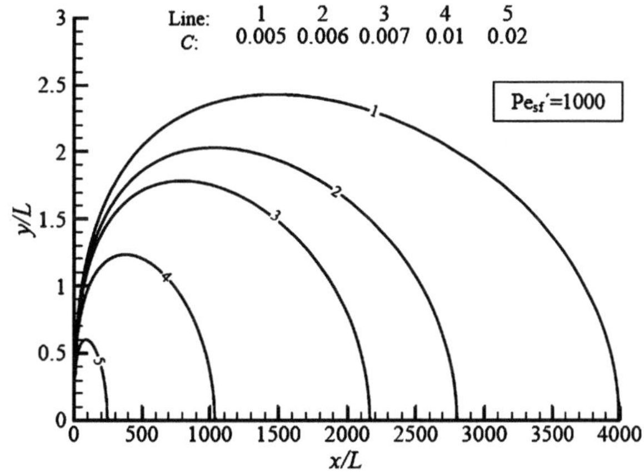

Fig. 7.1 Concentration profiles obtained with Eq. 7.3, at long distances from the flat slab

Making $D_T \cong D_L \cong D'_m$ a diffusion/convection cloud is again formed and the corresponding concentration contour plots are given by the solution for the "continuous point source" [100] of two-dimensional solute transport,

$$c = \frac{n/b}{2\varepsilon\pi D'_m} \exp\left[\frac{u_0 x}{2 D'_m}\right] K_0 \left\{ \sqrt{\frac{u_0^2 (x^2 + y^2)}{4 {D'_m}^2}} \right\} \tag{7.2}$$

where b is the depth and K_0 is the modified Bessel function of second kind and zero order. In our case, the problem of mass transfer from a soluble slab (with length L) immersed in a granular bed of inerts through which fluid flow with uniform interstitial velocity and with n from Eq. 7.1, one obtains after re-arrangement:

$$C = \frac{\sqrt{\frac{2}{\pi} {Pe'_{sf}}^{1/4} + \frac{4}{\pi} Pe'_{sf}}}{2\pi} \exp\left[\frac{Pe'_{sf}}{2}\frac{x}{L}\right] K_0 \left\{ \sqrt{\frac{{Pe'_{sf}}^2}{4}\left[\left(\frac{x}{L}\right)^2 + \left(\frac{y}{L}\right)^2\right]} \right\} \tag{7.3}$$

This equation may be used to estimate the region of propagation of a given concentration level.

Figure 7.1 show the concentration contour plots obtained, taking $Pe' = 1{,}000$ as an example, for low values of C. As the value of C decreases, the distance of the contour surfaces to the solid soluble particle increase and the solution for the "continuous point source" approach to the "exact" solution, obtained numerically. This result was confirmed by Gurdes de Carvalho et al. [31], who analysed the accuracy of the "continuous point source" solution, for different ranges of C, in the case of an active sphere immersed in a packed bed. The authors showed that for the higher values of C the concentration contours obtained with "continuous point source" solution deviates considerably from the "exact" solution, obtained numerically. However, for low values of C and if a correct value of n is used, true coincidence is observed between the numerical results and the "continuous point source" solution.

7.1.2 Concentration Profiles From a Cylinder Aligned With Flow

The analytical solution for a continuous point source has also been derived by Wexler [100], solving the three-dimensional solute-transport equation from a point source. The solution is given by

$$c = \frac{n}{4\varepsilon\pi(x^2+y^2)^{1/2}D'_{\mathrm{m}}}\exp\left[\frac{u_0\left[x-(x^2+y^2)^{1/2}\right]}{2D'_{\mathrm{m}}}\right]. \tag{7.4}$$

with $D_{\mathrm{T}} \cong D_{\mathrm{L}} \cong D'_{\mathrm{m}}$ and a good estimate for n is required. The point source is located at the point $(x, y) = (0, 0)$. In our case, the problem of mass transfer around a soluble cylinder (with length L and diameter d_1) immersed in a granular bed of inerts through which fluid flow with uniform interstitial velocity, the mass flux rate is expressed as,

$$n = \varepsilon k\pi d_1 L(c^* - c_0) = \mathrm{Sh}'_{\mathrm{c}}\frac{D'_{\mathrm{m}}}{\mathrm{L}}\varepsilon\pi d_1 L(c^* - c_0) \tag{7.5}$$

with Sherwood number, $\mathrm{Sh}'_{\mathrm{c}}$, given by Eq. 6.73. Making use of the dimensionless variables (please see Eq. 6.58), Eq. 7.4 results, after re-arrangement:

$$C = \frac{\sqrt{\frac{2}{\pi}\mathrm{Pe}_{\mathrm{c}}'^{1/4} + \frac{4}{\pi}\mathrm{Pe}'_{\mathrm{c}} + \left(6\frac{L}{d_1}\right)^{1/2} + \left(\frac{4}{3}\frac{L}{d_1}\right)^{3/2} + \frac{5}{3}\mathrm{Pe}_{\mathrm{c}}'^{5/9}\left(\frac{L}{d_1}\right) - \mathrm{Pe}_{\mathrm{c}}'^{2/9}\left(2\frac{L}{d_1}\right)^{1/3}}}{4(L/d_1)\left[(x/L)^2 + (y/L)^2\right]^{1/2}}$$
$$\times\exp\left\{\frac{\mathrm{Pe}'_{\mathrm{c}}}{2}\left[x/L - \left[(x/L)^2 + (y/L)^2 +\right]^{1/2}\right]\right\} \tag{7.6}$$

Figure 7.2 show the concentration contour plots obtained, taking $\mathrm{Pe}'_{\mathrm{c}} = 1{,}000$, as an example, for low values of C.

7.1.3 Concentration Profiles From a Soluble Sphere

In this model situation, we investigate the steady state conditions when a soluble sphere releases solute to the flowing fluid, with uniform interstitial velocity u_0, at a constant rate, n. Again, it is assumed that the volume of additional fluid injected into the main fluid, with the solute, is negligible.

A diffusion/convection cloud is formed and the corresponding concentration contour plots for the "continuous point source", at the point $(x = 0,\ y = 0)$, are also given by Eq. 7.4. If n from Eq. 6.82 is substituted in Eq. 7.4, the resulting equation may be re-arranged to read

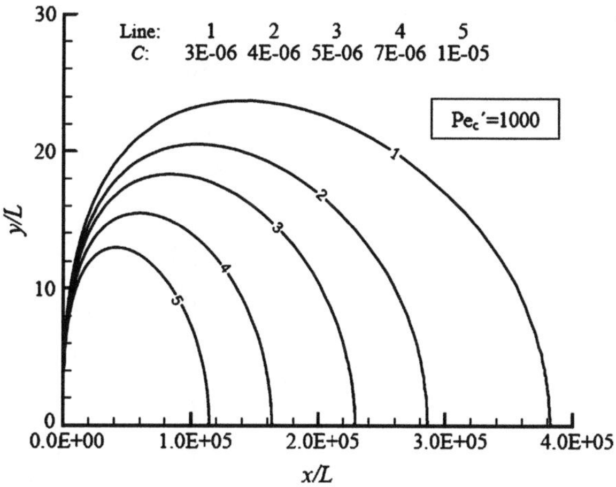

Fig. 7.2 Concentration profiles obtained with Eq. 7.6, at long distances from the cylinder

$$C = \frac{\left[4 + \frac{4}{5}\mathrm{Pe}_s'^{\,2/3} + \frac{4}{\pi}\mathrm{Pe}_s'\right]^{1/2}}{2\left[(x/a)^2 + (y/a)^2\right]^{1/2}} \exp\left\{\frac{\mathrm{Pe}_s'}{4}\left[(x/a) - \left[(x/a)^2 + (y/a)^2\right]^{1/2}\right]\right\} \tag{7.7}$$

Figure 7.3 show the concentration contour plots obtained, taking $\mathrm{Pe}_s' = 1,000$, as an example, for low values of C.

7.1.4 Concentration Profiles From a Cylinder in Cross Flow

If a slightly soluble cylinder, buried in a packed bed, is exposed to uniform fluid flow with velocity u_0, it will then release solute at a rate n, given by Eq. 6.91.

A diffusion/convection cloud is formed and the corresponding concentration contour plots for the "continuous point source", at the point ($x = 0$, $y = 0$), are also given by Eq. 7.4, and the substitution of n/b from Eq. 6.91 into 7.4 gives then

$$C = \frac{\left[\frac{2}{\pi^2}\mathrm{Pe}_{cf}'^{\,1/4} + \frac{32}{\pi^3}\mathrm{Pe}_{cf}'\right]^{1/2}}{2} \exp\left[\frac{\mathrm{Pe}_{cf}'}{4}\frac{x}{a}\right] \times K_0\left\{\sqrt{\frac{\mathrm{Pe}_{cf}'^{\,2}}{16}\left[\left(\frac{x}{a}\right)^2 + \left(\frac{y}{a}\right)^2\right]}\right\} \tag{7.8}$$

Concentration contours given by Eq. 7.8 are shown in Fig. 7.4, for $\mathrm{Pe}_{cf}' = 1,000$ (note that for $d_1 = 2$ m, $u_0 = 5 \times 10^{-4}$ mm/s and $D_m' \cong 10^{-9}$ m^2/s, there results $\mathrm{Pe}_{cf}' \cong 1,000$). From a detailed study of a similar nature, for flow past a sphere, it is reasonable to expect that values of the downstream reach of each contour surface (defined as the value of x/a, for $y/a = 0$, for that surface),

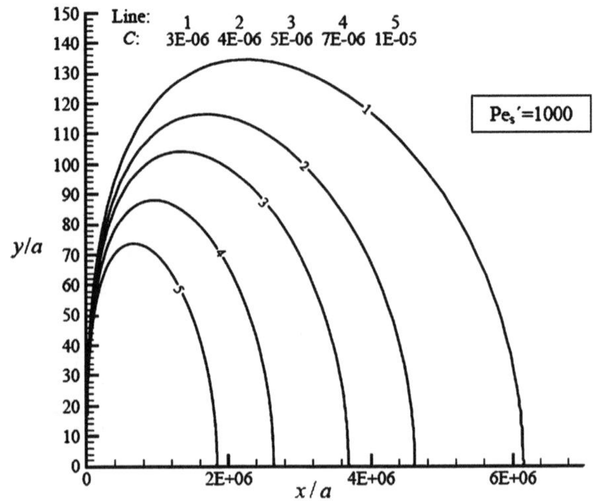

Fig. 7.3 Concentration profiles obtained with Eq. 7.7, at long distances from the sphere

predicted by Eq. 7.8, will differ by less than 10% from the exact value, if $\mathrm{Pe}' > 500$ and $C < 0.05$. The accuracy of the prediction will improve as C is decreased.

7.1.5 Concentration Profiles From a Prolate Spheroid

If a soluble prolate spheroid, buried in a packed bed, is exposed to uniform fluid flow with uniform interstitial velocity u_0, it will then release solute at a rate n, given by

$$n_\mathrm{T} = \varepsilon D'_\mathrm{m}\sqrt{\frac{4}{\pi}\mathrm{Pe}'_\mathrm{ps}}\left(\frac{2}{3}\frac{e^3}{e-(1-e^2)\tanh^{-1}e}\right)^{1/2}\frac{2\pi a}{(1-e^2)^{1/6}}(C^*-C_0) \qquad (7.9)$$

with Sherwood number, Sh'_ps, given by Eq. 6.115. Making use of the dimensionless variables, Eq. 7.4 results, after re-arrangement:

$$\frac{C-C_0}{C^*-C_0} = \frac{\frac{1}{2}\left[\frac{4}{\pi}\mathrm{Pe}'_\mathrm{ps}\right]^{1/2}\left(\frac{2}{3}\frac{e^3}{e-(1-e^2)\tanh^{-1}e}\right)^{1/2}}{4\left[(x/d_\mathrm{eq})^2+(y/d_\mathrm{eq})^2\right]^{1/2}} \times \exp\left\{\frac{\mathrm{Pe}'_\mathrm{ps}}{2}\left[x/d_\mathrm{eq}-\left[(x/d_\mathrm{eq})^2+(y/d_\mathrm{eq})^2\right]^{1/2}\right]\right\} \qquad (7.10)$$

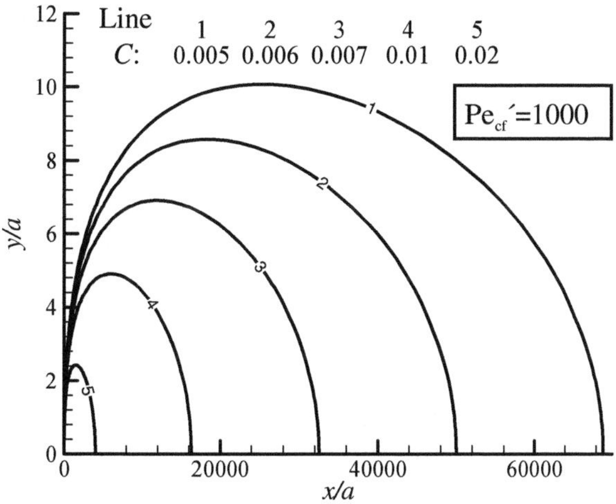

Fig. 7.4 Concentration profiles obtained with Eq. 7.8, at long distances from the cylinder in cross flow

7.1.6 Concentration Profiles From an Oblate Spheroid

If a slightly oblate spheroid, buried in a packed bed, is exposed to uniform fluid flow with uniform interstitial velocity u_0, it will then release solute at a rate n, given by

$$n_T = \varepsilon D'_m \sqrt{\frac{4}{\pi}\mathrm{Pe}'} \left(\frac{4}{3} \frac{e^3}{-e + e^3 + \sqrt{1-e^2}\cot^{-1}(\sqrt{1/e^2 - 1})} \right)^{1/2} \times \frac{2\pi a^2 (1-e^2)^{1/3}}{d_{eq}} (C^* - C_0) \tag{7.11}$$

with Sherwood number, Sh'_{os}, given by Eq. 6.129. Making use of the dimensionless variables, Eq. 7.4 results, after re-arrangement:

$$\frac{C - C_0}{C* - C_0} = \frac{\frac{1}{2}\left[\frac{4}{\pi}\mathrm{Pe}'_{os}\right]^{1/2} \left(\frac{4}{3} \frac{e^3}{-e + e^3 + \sqrt{1-e^2}\cot^{-1}(\sqrt{1/e^2 - 1})} \right)^{1/2}}{4\left[(x/d_{eq})^2 + (y/d_{eq})^2\right]^{1/2}} \times \exp\left\{ \frac{\mathrm{Pe}'_{os}}{2} \left[x/d_{eq} - \left[(x/d_{eq})^2 + (y/d_{eq})^2\right]^{1/2} \right] \right\} \tag{7.12}$$

Figure 7.5 show the concentration contour plots obtained, taking $\mathrm{Pe}'_{os} = 1{,}000$ and $e = 0.5$ as an example, for low values of concentration, C. As the value of C decreases, the distance of the contour surfaces to the solid soluble particle increase and the solution for the "continuous point source" approach to the "exact" solution, possible to obtain numerically (i.e. if a correct value of n is used, true coincidence is observed).

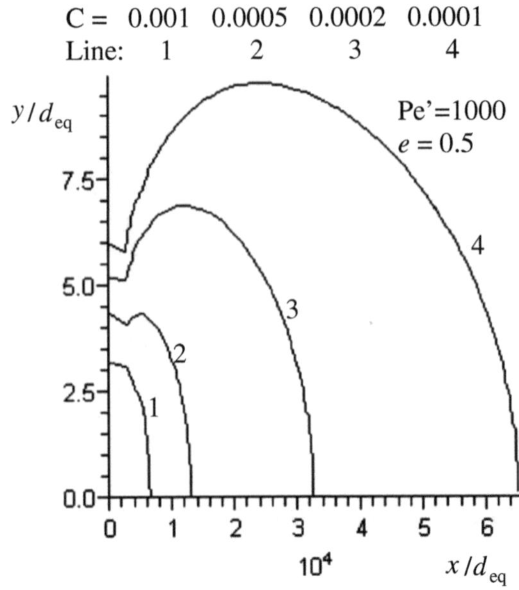

Fig. 7.5 Concentration profiles obtained with Eq. 7.12, at long distances from the oblate spheroid

The problem of mass transfer between a buried solid with different shapes and the fluid flowing along it in a granular bed, lends itself to a simple full theoretical analysis, under an appropriate set of conditions.

The elliptic partial differential equations resulting from a differential mass balance have been solved numerically over a wide range of values of the relevant parameters. It was found that the resulting Sherwood values are in good agreement with experimental results.

Results of the exact numerical solutions and analytical solutions were also used to predict the solute migration from an active solid particle buried in a packed bed of inert particles, through which fluid flows with uniform velocity. The concentration contour surfaces were obtained using an analytical solution of continuous injection of solute at a point source in a uniform stream and the proposed correlations for the mass transfer rate developed numerically or analytically.

7.2 Rising Damp in Building Walls

Rising damp, a world-wide phenomenon, is a major cause of decay in masonry materials such as stone, brick and mortar. The conservation of historic buildings has become important nowadays and has developed significantly in recent years. Rising damp in historic buildings may be considered one of the most important of all the different manifestations of dampness, leading to the destruction of stone materials due to frost/defrost cycles and the presence of salts associated with crystallization (Künzel [61] argues that in many cases increased salt content is interpreted as rising damp), decay in elements in contact with stone walls, such as wooden beams, wooden ceilings and other finishings, and insanitary conditions caused by excess humidity associated with the development of fungus and mould.

Rising damp, which originates in the ground and permeates porous materials by capillarity, is one of the main causes of decay in old buildings, particularly those containing thick walls built from different materials. Having identified the particular characteristics of those buildings, it is important to recognize the limitations of traditional technologies and to investigate new solutions for the phenomenon. The traditional techniques used to deal with this kind of problem (such as watertight barriers, the injection of hydrofuge products, etc.) sometimes prove ineffective or are very expensive, justifying the need to find new approaches (see [34]).

The Building Physics Laboratory (LFC), at the University of Porto-Faculty of Engineering (FEUP) has been developing important experimental research into the problems of rising damp. In recent years the principles for a treatment technique called the "Wall Base Ventilation System to Treat Rising Damp–HUMIVENT device" has been validated and characterised in experiments. The technique consists of circulating air around the base of very thick walls built from different materials, with a relative humidity saturation distance. Wall base ventilation increases evaporation, which reduces the level of the damp front. This is possible only when the groundwater is lower than the base of the wall [17].

In previous research it was observed that the "wall base ventilation device" reduces the level of the wet front. Following this, a series of experiments were carried out to explore boundary conditions, geometry, engine speed, etc., and it was possible to observe that a hygro-regulated system is essential in controlling possible condensation inside the system. A hygro-regulated system is a mechanical system controlled by probes placed at the inlet and at the outlet. In accordance with pre-programmed criteria, the probes will switch the system on/off.

7.2.1 Rising Damp Theory

The phenomenon of rising damp can be analysed using the sharp front model developed by Hall and Hoff [39]. This model describes the relationship between sorptivity, S, wall thickness, b, evaporation potential, e, and capillary rise heights, h.

Considering the absorption inflow and evaporation loss to be balanced and disregarding gravitational forces (see Fig. 7.6), the following differential equation is obtained [40]

$$b\frac{di}{dt} = b\frac{S}{2}t^{-1/2} - eh \tag{7.13}$$

where i is the volume of liquid absorbed per unit cross section in time t [36]. Considering that $i = \theta_{\mathrm{w}}h$, the following expression is obtained

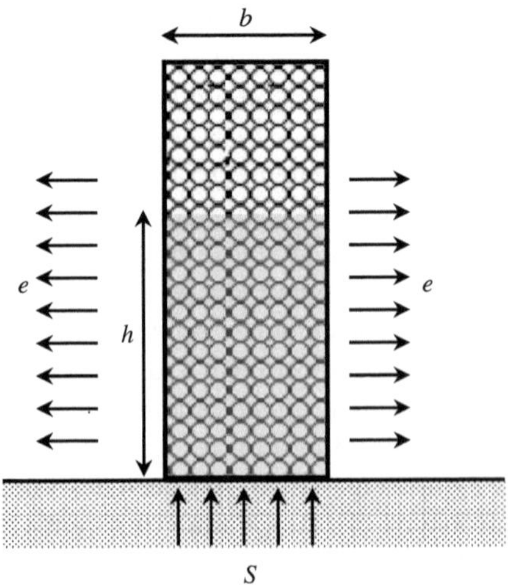

Fig. 7.6 Sketch of sharp front model

$$\frac{dh}{dt} = \frac{S}{2\theta_{\mathrm{w}}} t^{-1/2} - \frac{eh}{b\theta_{\mathrm{w}}} \tag{7.14}$$

with $i = St^{1/2} = \theta_{\mathrm{w}} h$, results

$$\frac{dh}{dt} = \frac{S^2}{2\theta_{\mathrm{w}}^2} \frac{1}{h} - \frac{eh}{b\theta_{\mathrm{w}}} \tag{7.15}$$

The analytical solution of Eq. 7.15 is

$$h = S\sqrt{\frac{b}{2e\theta_{\mathrm{w}}}\left[1 - \exp\left(-\frac{2e}{b\theta_{\mathrm{w}}} t\right)\right]} \tag{7.16}$$

with the equilibrium height, h_∞, given by

$$h_\infty = S\sqrt{\frac{b}{2e\theta_{\mathrm{w}}}} \tag{7.17}$$

7.2.2 The Wall Base Ventilation System

The wall base ventilation system (HUMIVENT device) consists of ventilating the base of walls using a natural ventilation process or by installing a hygro-regulated mechanical ventilation device (see Fig. 7.7). In terms of analysis, we considered a wall base ventilation system along which air is flowing close to the "saturated" wall $(0 < z < L)$.

In laminar flow, the transport of moisture and temperature across the boundary layers is controlled by molecular diffusion. The Reynolds number, Re, may be interpreted as the ratio of the flow of destabilising forces to stabilising forces

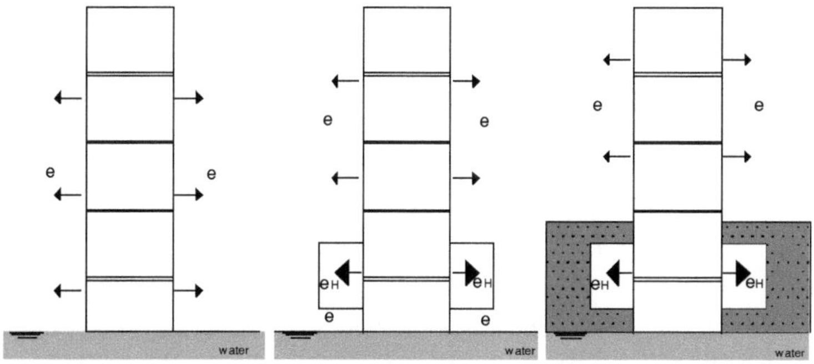

Fig. 7.7 Diagram of the different boundary conditions studied

(viscosity). Stable laminar flow is thus characterized by low Re values. Laminar flow occurs at Reynolds numbers of less than approximately 3×10^5. In our study $Re < 10^5$, in all experiments.

If the analysis is restricted to situations in which the moisture transfer boundary layer is thin and if a small control volume is considered inside this boundary layer, with side lengths δz, δy and unity (perpendicular to the plane of the figure), a steady state material balance in the solute leads to

$$u\frac{\partial c}{\partial z} = D_m \frac{\partial^2 c}{\partial y^2} + D_m \frac{\partial^2 c}{\partial z^2} \tag{7.18}$$

where D_m is the molecular diffusion coefficient, in the cross stream and in the stream wise directions. If the boundary layer is thin compared to the length of the saturated wall, the last term on the right-hand side of Eq. 7.18 is likely to be negligible. For the situation outlined in Fig. 7.6, the boundary conditions are,

$$c = c_0 \quad z = 0 \quad y > 0 \tag{7.19a}$$

$$c = c^* \quad z > 0 \quad y = 0 \tag{7.19b}$$

$$c \to c_0 \quad z > 0 \quad y \to \infty \tag{7.19c}$$

and the analytical solution is given by [21]

$$\frac{c - c_0}{c^* - c_0} = \text{erfc}\left(\frac{y}{2\sqrt{D_m z/u}}\right) \tag{7.20}$$

where c_0 is the bulk concentration of water vapour, c^* is the equilibrium concentration of water vapour, and u is the air velocity. The flux of evaporation, N, at the saturated wall surface may be obtained from (7.20) as

$$N = -D_m \left(\frac{\partial c}{\partial y} \right)_{y=0} = (c^* - c_0) \left(\frac{D_m}{\pi z/u} \right)^{1/2} \tag{7.21}$$

Finally, it is possible to obtain the total rate of solid evaporation, n, over the whole wall surface

$$n = \int_0^L N h_{\mathrm{H}} \, dx = (c^* - c_0) h_{\mathrm{H}} L \left(\frac{4D_m}{\pi L/u} \right)^{1/2} \tag{7.22}$$

where $N = -D_m(\partial c/\partial y)_{y=0}$ and h_H is the width of HUMIVENT system.

If we consider a rising damp steady state, involving a wall buried in sand and a wall base ventilation system (HUMIVENT) located on both sides above the base (see Fig. 7.7), Eq. 7.17 produces

$$\frac{bS^2}{2\theta_{\mathrm{w}} h_\infty} = e(h_\infty - h_s - h_H) + e_H h_H + e_s h_s \tag{7.23}$$

with e_H (related to HUMIVENT) given by [34]

$$e_H = \frac{n}{L.h_H \rho_w} = \frac{(c^* - c_0)}{\rho_w} \left(\frac{4D_{\mathrm{m}}}{\pi L/u} \right)^{1/2} \tag{7.24}$$

and the steady-state height of rise given by

$$h_\infty = \left(1 - \frac{e_s}{e}\right) \frac{h_s}{2} + \left(1 - \frac{e_H}{e}\right) \frac{h_H}{2} + \sqrt{\frac{bS^2}{2\theta_{\mathrm{w}} e} + \left(\frac{e_s}{e} - 1\right)^2 \frac{h_s^2}{4} + \left(\frac{e_H}{e} - 1\right)^2 \frac{h_H^2}{4} + \left(1 - \frac{e_s}{e} - \frac{e_H}{e} + \frac{e_s e_H}{e^2}\right) \frac{h_s h_H}{2}} \tag{7.25}$$

Noting that the evaporation potential related with sand placed on both sides of the wall above its base, e_s, will be negligible (sand water saturated) in comparison with the other evaporation rates, e and e_H, and Eq. 7.25 reduces then to

$$h_\infty = \frac{h_s}{2} + \left(1 - \frac{e_H}{e}\right) \frac{h_H}{2} + \sqrt{\frac{bS^2}{2\theta_{\mathrm{w}} e} + \frac{h_s^2}{4} + \left(\frac{e_H}{e} - 1\right)^2 \frac{h_H^2}{4} + \left(1 - \frac{e_H}{e}\right) \frac{h_s h_H}{2}} \tag{7.26}$$

and for $e_H = e_s = 0$, Eq. 7.26 reduces to

$$h_\infty = \frac{h_s}{2} + \frac{h_H}{2} + \sqrt{\frac{bS^2}{2\theta_{\mathrm{w}} e} + \frac{h_s^2}{4} + \frac{h_H^2}{4} + \frac{h_s h_H}{2}} \tag{7.27}$$

Finally, to estimate the evaporation potential, *e*, many existing formulae can be applied [53]. In this study we used the Penman–Monteith equation [34] which, given that the latitude of Porto is 40.15°, results in an evaporation potential of 0.0017 mm/min.

7.2.3 Numerical Simulation

The simulation programs to evaluate changes in the moisture content and temperature inside walls are essential instruments for simulating the wall's behaviour in the presence of moisture, depending on the internal and external climatic conditions [12].

The calculation program used in the numerical simulations for the experimental and analytical validations was WUFI-2D, developed by Fraunhofer Institute for Building Physics. The governing equation for moisture transport is [44]

$$\frac{dw}{d\phi}\frac{\partial \phi}{\partial t} = \nabla(D_\varphi \nabla \phi + \delta_p \nabla \phi p_s) \tag{7.28}$$

and the governing equation for heat transport is

$$\frac{dH}{dT}\frac{\partial T}{\partial t} = \nabla(\lambda_T \nabla T) + h_v \nabla(\delta_p \nabla \varphi p_s) \tag{7.29}$$

where dH/dT is the heat storage capacity of the moist building material, $dw/d\phi$ is the moisture storage capacity, λ_T is the thermal conductivity, D_ϕ is the liquid conduction coefficient, δ_p is the water vapour permeability, h_v is the evaporation enthalpy of the water, p_s is the water vapour saturation pressure, T is the temperature and ϕ is the relative humidity.

As hygrothermal modelling offers a powerful tool for predicting heat and moisture transport through multi-layer building assemblies. In this work, the hygrothermal model was used to compare the results of three case studies (monolithic wall) under different natural conditions (see Fig. 7.7). The laboratory climate was modelled by a sine wave varying between 19°C, 50% RH in winter and 21°C, 70% RH in summer, which corresponds to a normal moisture load (mean values of 60% RH and temperature of 20°C). The initial moisture content of the used materials was the equilibrium moisture at 80% RH. The simulation periods began on January 1.

The vapour diffusion thickness value used was zero (no coating) and the interior heat transfer coefficient was constant and equal to 8 W/m^{2}K. The exterior heat transfer coefficient only contained the convective part and was considered independent from the wind (constant value of 17 W/m^{2}K).

7.2.4 Rising Damp Analysis

7.2.4.1 In the Laboratory

The physical model selected consisted of a prismatic wall 1.58 m high, 2.00 m long and 0.20 m thick, waterproofed on the two upper sides (measuring

Table 7.1 Properties of the limestone sample [93]

		Limestone	
Bulk density, ρ (kg/m^3)		2155 ± 14	
Heat capacity, c_p (J/kgK)		1,000	
Porosity, ε (%)		19.7	
Thermal conductivity, λ_T (W/mK)		1.33 ± 0.025	
Vapor diffusion resistance factor, μ (−)	Dry cup:	41 ± 2.2	
	Wet cup:	29 ± 3.1	
Moisture storage function, w (kg/m^3)	HR (%)	*Adsorption*	*Desadsorption*
	4.0	0.521	0.951
	11.2	0.593	1.150
	34.8	0.872	1.239
	58.6	1.043	1.628
	76.3	1.237	2.360
	80.0	1.334	–
	84.2	1.584	–
	92.1	2.381	4.144
Capillary transport coefficient, D_w (m^2/s)	$w_{80\%} = 1.7$ kg/m^3	6.6×10^{-11}	
	$w_{sat} = 188$ kg/m^3	6.2×10^{-8}	
Water absorption coefficient, A (kg/m^{2}s$^{1/2}$)		0.024 ± 0.002	
Free-water saturation, w_f (kg/m^3)		177 ± 1.8	

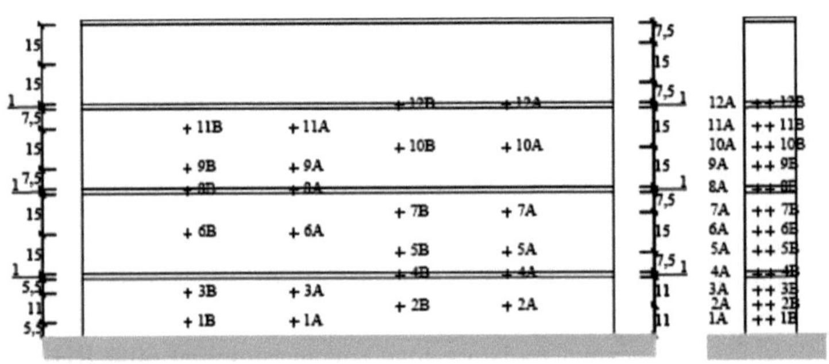

Fig. 7.8 Position of the relative humidity and temperature probes [93]

1.58 × 0.20 m^2) to prevent moisture entering from this direction. The limestone walls (see Table 7.1) were placed in a reservoir approximately 2.20 × 2.50 × 0.50 m^3 made from cement blocks. To assess moisture transfer inside the walls, probes were inserted at different heights and depths to measure relative humidity and temperature, as sketched in Fig. 7.8. These probes were then connected to a data acquisition and recording system.

The configurations tested are shown in Fig. 7.7. In Configuration 1 the base of the wall is immersed up to a height of 8 cm and, as we wished to assess the effect of the wall base ventilation system–HUMIVENT, in Configurations 2 and 3, a ventilation box was placed on both sides of the wall. The only difference between these configurations was that in Configuration 3 we measured the behaviour of a

Table 7.2 Experi ment values (average) for relative humidity and temperature inlet in the wal l and system outlet

Configuration 1	Configuration 2		Configuration 3	
ϕ_O (%)	ϕ_O (%)	ϕ_s (%)	ϕ_O (%)	ϕ_s (%)
60.0	57.8	60.1	58.2	60.3
T_O (°C)	T_O (°C)	T_s (°C)	T_O (°C)	T_s (°C)
20.0	22.6	22.4	23.0	22.8

Table 7.3 Comparison between experimental, analytical and numerical results (obtained with an error lesser than 5%)

h_∞ (mm)	Limestone		
	Conf. 1	Conf. 2	Conf. 3
Experiments	455–540	380–455	455–540
WUFI 2D (ϕ)	550	–	490
WUFI 2D (w)	520	–	490
Analytical solution	510	380	451
ϕ_{eq} (%)	60		
w_{eq} (kg/m^3)	1.047		

wall with both sides buried in sand to a height of 45 cm. The sand was saturated (100% RH) during the tests. To avoid turbulence, an air velocity of 0.083 m/s was used and the relative humidity and temperature values for the experiments are presented in Table 7.2.

The climatic conditions inside the laboratory during the experiments were considered constant at a temperature of 23°C and a relative humidity of 58%. Under these conditions and considering an air velocity inside the laboratory of 1 cm/s, the evaporation potential was given by [87]

$$e = D_m \frac{(c^* - c_0)}{\rho_w} 0.646 \frac{\mathrm{Re}^{1/2}\,\mathrm{Sc}^{1/3}}{L} \tag{7.30}$$

The resulting evaporation potential value is 0.46×10^{-4} mm/min. The evaporation potential associated with the HUMIVENT system, e_H, is given in Eq. 7.24 as 2.17×10^{-4} mm/min. If these values are inserted into Eq. 7.25, a steady state height of rise of 379.5 mm is obtained for Configuration 2 and a steady flow rate through the wall of about 27.1 L/year per metre length of wall. This value is very similar to the experimental results obtained, of between 380 and 455 mm (as sketched in Fig. 7.8, position 6 and 7–average value of 417.5 mm), and significantly different from the steady state height of rise obtained without the use of the HUMIVENT system, which was $h_\infty = 540$ mm in Configuration 1 (experimental value between 455 and 540 mm—Fig. 7.8, position 7 and 8—as shown in Table 7.3).

In the analyse of our experimental results, the following expression was used for water vapour diffusivity in air [56]

$$D_{\mathrm{m}}(m^2/s) = 2.23 \times 10^{-5} \left[\frac{273.15 + T(^\circ C)}{273.15}\right]^{1.5} \tag{7.31}$$

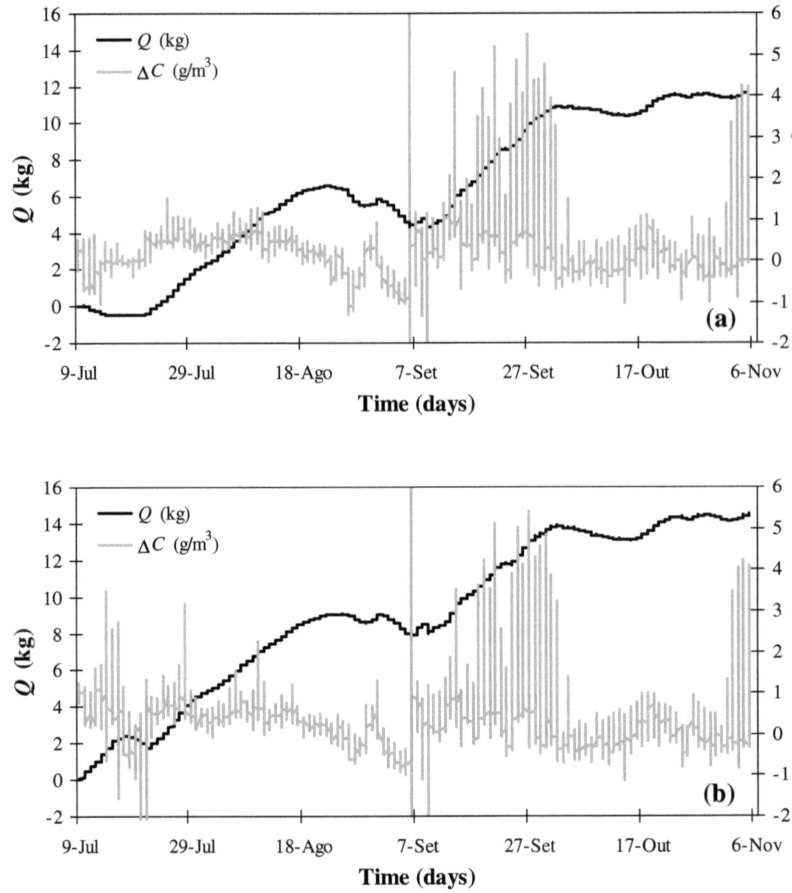

Fig. 7.9 Quantity of water vapour extracted and concentration differential (**a**) in Configuration 2 and (**b**) in Configuration 3

Table 7.3 shows the experimental results for the first series of laboratory tests, i.e. the rising damp values for the steady height of rise under different boundary conditions. The results of the experiment show that the presence of a wall base ventilation system on both sides prevents the damp front (see results) and that the simple analytical model proposed by Hall and Hoff [39] clearly describes the rising damp front when compared with the experiment results. During the four months of experimental research, in laboratory tests on Configurations 2 and 3 the new treatment technique extracted approximately 11.8 and 14.5 kg of water respectively (see Fig. 7.9a, b).

The simulation program provided the water content, the relative humidity and the temperature at all the points of the pre-defined grid and their evolution for the duration of the simulation. It also gave the heat and moisture flows along its

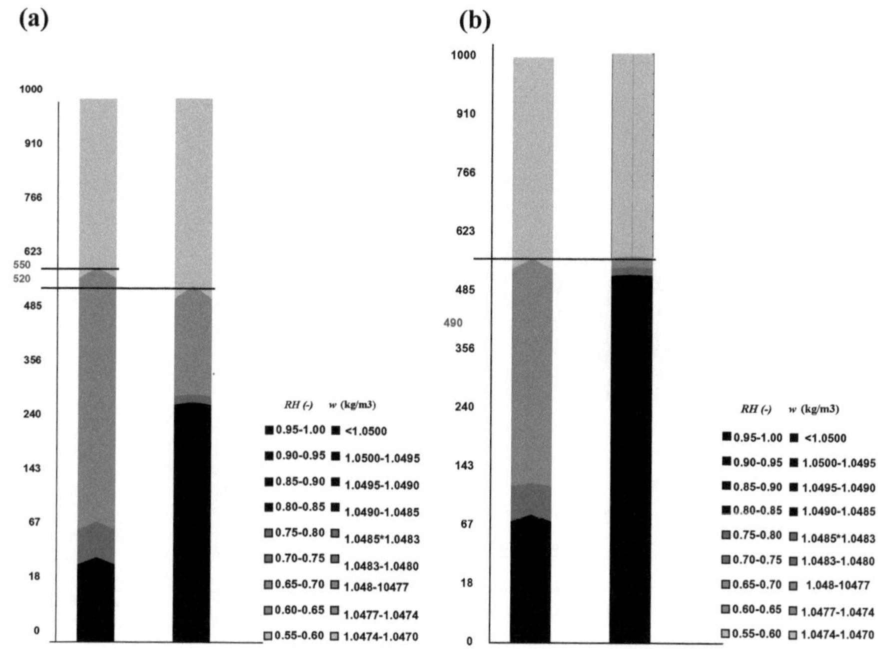

Fig. 7.10 Variation of rising damp front with relative humidity and water content for (**a**) Configuration 1 and (**b**) Configuration 3

surfaces. Figure 7.10a, b show the variation of rising damp front with relative humidity and water content for different boundary conditions. These figures also show that the numerical results are in accordance with the experimental and analytical values obtained (see Table 7.3).

The research group is currently studying whether these values may have adverse consequences in terms of the long-term deterioration of the wall materials [99]. As we understand how salts dissolve in water and their behaviour in terms of crystallization/dissolution, the target, in future, will be to scale the new rising damp treatment system to avoid crystallization/dissolution problems.

7.2.4.2 In-Field

The hygro-regulable system was installed in a church in Northern Portugal (see Fig. 7.11a). Inside the building, two hygro-regulable mechanical ventilation subsystems were installed (see Fig. 7.11b). In the Southside subsystem, air was admitted through grids located inside the building, and was extracted into the cloister. Extraction was controlled by a hygro-regulable engine of variable speed [26]. The system had two probes for measuring relative humidity and

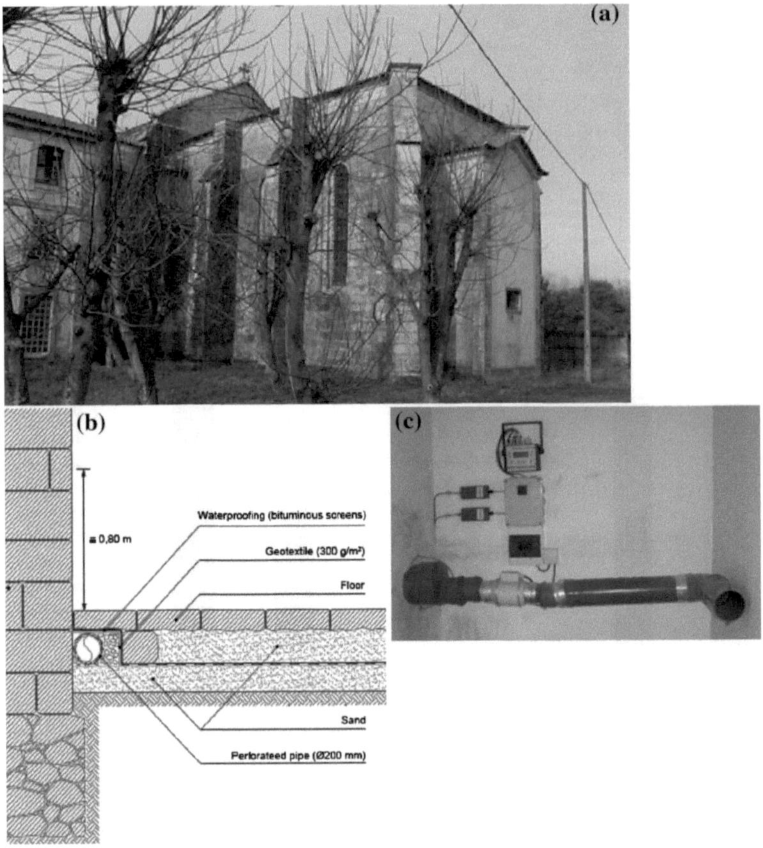

Fig. 7.11 (**a**) Church in Northern Portugal; (**b**) Wall base ventilation system and (**c**) Data acquisition and recording system

temperature, two transmitters, a control module and a data acquisition system for recording results (see Fig. 7.11c).

The system installed in the church initially began operating whenever the relative humidity at the exit was 5% higher than the relative humidity at the entrance. The idea was to admit dry air comparing to the air inside the system. This criterion was found to be inadequate, as it meant that the system was operating at periods when condensation occurred inside it. Consequently, a new criterion was proposed with a view to optimizing the system, based upon the difference in vapour pressure (ΔP) at the exit and entrance. The system now began functioning whenever the ΔP was positive.

Figure 7.12a, b shows the functioning periods of the ventilator, in a north and south wall, and the importance of HUMIVENT device in rising damp treatment, by the quantity of water vapour extracted during the five months analyzed (272.3 kg in the north wall and 170.2 kg in the south wall).

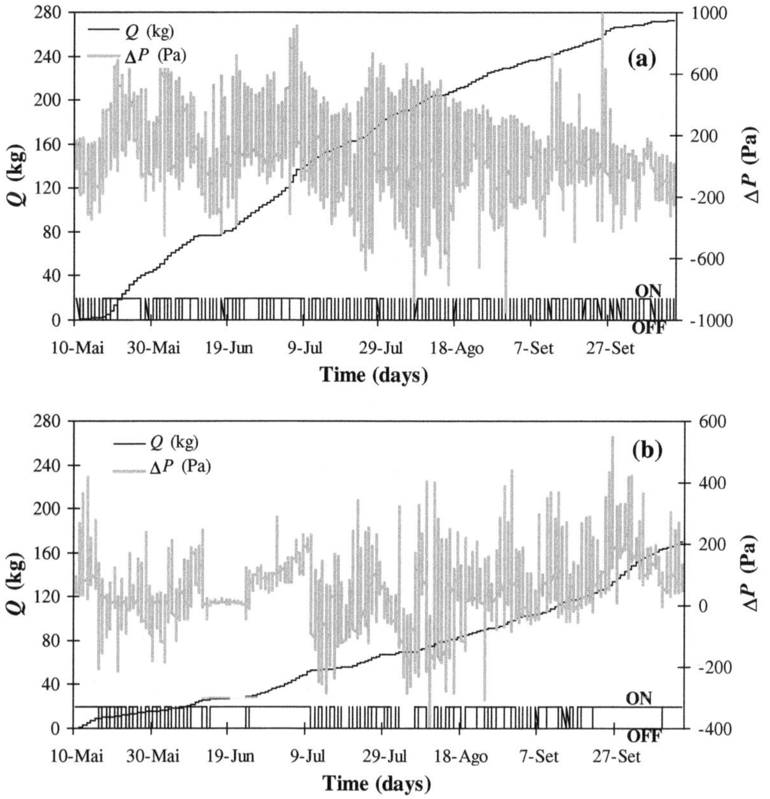

Fig. 7.12 Quantity of water vapour extracted, pressure differential and functioning of system: (**a**) North wall and (**b**) South wall

The entry of air with very low relative humidity could generate the crystallization of salts existing in the building materials, threatening its durability. For this reason, the relative humidity value at the entrance had to be limited. The relative humidity scores recorded, which range from 60 to 95%, are not considered to present a risk of salt crystallization/dissolution inside the system, consider the salts detected in those area. However, the problem might arise in another type of external climate or with other salts type [99]. In that case the HUMIVENT device must also limit the inlet relative humidity value.

In conclusion, the mechanisms of moisture transfer are complex, particularly in terms of rising damp in historic buildings. As rising damp is one of the main causes of decay in these buildings, it is important to study the factors related to this phenomenon.

The traditional techniques currently used to minimise rising damp are not effective or too expensive, in particular when dealing with walls of considerable thickness and heterogeneous materials. Laboratory tests on 20 cm thick limestone walls show that an efficient way of treating rising damp in historic buildings is to ventilate the wall base using the HUMIVENT device.

The analytical model of rising damp predicts steady height of rise figures that are consistent with laboratory observations, which provide a good validation of the analysis. Using this model it is possible to scale the HUMIVENT system, optimise it and predict its behaviour. The numerical simulation results are similar to those obtained experimentally and analytically, which provide a qualitative validation of the results obtained.

7.3 Bubbles and Drops in Porous Structures

Transport of mixtures is encountered in many industrial processes and engineering applications. More specifically, one of the most interesting cases is that of a discontinuous phase (droplets) situated within a flowing continuous aqueous phase. A typical example of this may be found in the oil industry, and is related to unconventional behavior found in some heavy oil reservoirs where a disconnected gas phase (gas bubbles) inside a connected liquid phase (heavy oil) is observed over periods of time long enough to consider that such a phase distribution governs flow dynamics [29, 42]. Qualitatively speaking, the aspects of such a transport process are rather well understood. In general, droplets are transferred by convection and diffusion while their population is strictly governed by local hydrodynamic phenomena [85]. It is very difficult to obtain a globally valid description for the consequent transient (dynamic) behavior of the microstructure of such a mixture due to the high geometrical complexity of such a configuration. For example, the domain size could grow because the domains may coalesce under flow conditions and a break-up process sets in when they are large enough to be affected.

Renewed interest in this field originates from theoretical investigations on the flow in porous materials using conventional computations [30, 73, 89], the Lattice Boltzmann Method [46, 68, 69] and pore network simulators [13, 24]. These approaches focus mainly on flow dynamics for multiphase and multi-component fluids rather than the microstructure formation where limited research has been done [59, 60, 64, 70, 98].

The exemplar application presented here aims to describe this micro-structure especially for mixtures flowing in a homogeneous porous medium under laminar conditions. In particular, the scope of this theoretical study is to investigate the physical/geometrical parameters and the flow conditions which could assure the mixture's homogeneity (water/droplets, i.e. continuous/discontinuous phase) in terms of mean droplet size. Focus is on the domain sizes formed during this process and the approach taken here is the so-called '$S\gamma$ concept', i.e. the 'moments of distribution', for droplet size description [52], where an arbitrary number of moments of the domain size distribution is used to describe the microstructure. In the same study, the method was also described in relation to experiments on bubble coalescence. Subsequently, Klahn et al. [57] and Agterof et al. [3] used the same method to analyze experiments on the transient behavior of emulsion droplet break-up in an impeller flow, and to analyze the behavior of double emulsions. All these examples demonstrated the method's power, and will be explained in detail below.

In this section the $S\gamma$ method is extended to phenomena related to porous structures. The problem examined here is the laminar flow of a biphasic mixture in a homogeneous porous medium. The domain size for the two phases not only depends on the flow itself but also on the break-up and coalescence processes taking place because of the flow. The method's fundamental idea is that the evolution of the moments of a distribution can be analyzed using a transport equation including a convective term, which can be coupled to the local flow characteristics, which should not be affected by the presence of droplets. This assumption is valid only for droplet sizes small enough to follow the bulk flow and adequately assure the homogeneity of the flowing mixture (see Eq. 7.38) and relative discussion). The source terms in the transport equation describe the local phenomena (break-up and coalescence) that affect particle size distribution. The main advantage of this approach is that analytical expressions (including probability parameters) could be derived for the characteristic magnitudes, leading to successful simulations through common and rather simple CFD techniques.

7.3.1 Moments of Distribution (Sγ)

The domain (droplet) size distribution can be described by a collection of moments of the distribution as [3, 52, 57]:

$$S_\gamma = n\int_0^\infty d^\gamma\, P(d)\, \mathrm{d}d \tag{7.32}$$

where γ is an integer indicating the specific moment, n is the total number density, and *P(d)* is the droplet size distribution. Some of these moments can be related directly to physical properties of the dispersion. More precisely [3, 52]:

$$S_0 = n\int_0^\infty P(d)\, \mathrm{d}d = n \tag{7.33}$$

is the total number of droplets (number density),

$$S_2 = n\int_0^\infty d^2\, P(d)\, \mathrm{d}d \tag{7.34}$$

is the specific interfacial area, and

$$S_3 = n\int_0^\infty d^3\, P(d)\, \mathrm{d}d = \frac{6\varphi}{\pi} \tag{7.35}$$

is an expression of the volume fraction, φ.

It can be easily shown that for two different γ values, the corresponding $S\gamma$ functions can produce a useful characteristic diameter $d_{\gamma_1\gamma_2}$ according to the relation:

$$d_{\gamma_1\gamma_2} = \left(\frac{\int_0^\infty d^{\gamma_1}\, P(d)\, \mathrm{d}d}{\int_0^\infty d^{\gamma_2}\, P(d)\, \mathrm{d}d}\right)^{\frac{1}{\gamma_1-\gamma_2}} = \left(\frac{S_{\gamma_1}}{S_{\gamma_2}}\right)^{\frac{1}{\gamma_1-\gamma_2}} \quad \text{for } \gamma_1 \neq \gamma_2 \tag{7.36}$$

From geometrical and physical point of view, the most interesting $d_{\gamma_1\gamma_2}$ is the area-weighted mean droplet diameter, d_{32}, given as:

$$d_{32} = \frac{S_3}{S_2} = \frac{\int_0^\infty d[d^2\, P(d)]\, \mathrm{d}d}{\int_0^\infty d^2\, P(d)\, \mathrm{d}d} \tag{7.37}$$

It is worth noticing that this diameter is a spatially averaged size which here is considered to represent the actual droplet size.

The main advantage of the $S\gamma$ function is that it satisfies the transport equation [52]. Starting from the well-known population balance equation and following the mathematical route presented by Kamp et al. [52], the following transport equation can be derived:

$$\frac{\partial nP(d)}{\partial t} + \nabla \cdot (\underline{u}nP(d)) = G - E \tag{7.38}$$

where $\underline{u}$ is the local velocity vector and G, E are the generation and the extinction terms, respectively. These terms are strongly related to droplet creation (break-up) and death (coalescence).

After multiplying each term by d^γ and subsequently integrating over all particle sizes, the following equation can be obtained:

$$\frac{\partial}{\partial t}\left[\int_0^\infty d^\gamma nP(d)dd\right] + \nabla \cdot \left[n\int_0^\infty d^\gamma \underline{u}P(d)dd\right] = \int_0^\infty (G-E)d^\gamma\, dd \tag{7.39}$$

By reversing the order of differentiation and integration, Eq. 7.39 can also be written as a transport equation:

$$\frac{\partial S_\gamma}{\partial t} + \nabla \cdot (\underline{u}_\gamma S_\gamma) = h_i^{(\gamma)} \tag{7.40}$$

where:

$$\underline{u}_\gamma = \frac{\int_0^\infty P(d)\underline{u}(d)d^\gamma dd}{\int_0^\infty P(d)d^\gamma dd} \tag{7.41}$$

Expression (7.41) equals the local velocity vector, $\underline{u}$, if the droplets have the same velocity as the flowing continuous bulk phase, an assumption adopted here. It is assumed that the local flow field is independent of the droplets, thus their mean diameter is too small to affect the homogeneity of the flowing phase. Even for porosities close to unity, the mean droplet size must be lower than 10^{-6} m for creeping or laminar flow conditions [64] to satisfy the above assumption.

In Eq. 7.40, $h_i^{(\gamma)}$ is the source term related to droplet break-up and coalescence, expressed in terms of the moments of the distribution. For the majority of engineering applications, transport equation (7.41) can be simplified to:

$$\frac{\partial S_\gamma}{\partial t} + \underline{u} \cdot \nabla S_\gamma = h_i^{(\gamma)} \tag{7.42}$$

When the flow problem is uncoupled from the transport one, the velocity field $\underline{u}$ can be obtained from typical CFD flow simulations.

7.3.2 Determination of Source Terms

Since the governing phenomena in the fluid phase are the droplet break-up and coalescence processes, the source term of the previous equation can be expressed as:

$$h_i^{(\gamma)} = h_{br}^{(\gamma)} + h_{cl}^{(\gamma)} \tag{7.43}$$

where $h_{br}^{(\gamma)}$ and $h_{cl}^{(\gamma)}$ are the respective source terms that can be modeled explicitly.

Break-up occurs when droplet diameter becomes larger than a critical value, d_{cr}. This value is related to the critical capillary number as follows:

$$d_{cr} = \frac{2\,\sigma\, Ca_{cr}}{\eta_c\, \dot{\gamma}} \tag{7.44}$$

where Ca_{cr} is the critical capillary number, $\dot{\gamma}$ is the local velocity gradient, η_c is the viscosity of the continuous phase, and σ is the interfacial tension. This critical diameter also depends on viscosity ratio and flow type, as discussed by Grace [32] and Stone [90].

The break-up source term, in its generic form, can be written as [49]:

$$h_{br}^{(\gamma)} = \int_0^\infty K_{br}(d)\Delta S_\gamma^{br}(d) nP(d) dd \tag{7.45}$$

where K_{br} is the break-up rate, and $\Delta S_\gamma^{br}(d)$ is the change in $S\gamma$ due to a single break-up event in a droplet of size *d*. An expression for $\Delta S_\gamma^{br}(d)$ follows from the conservation of dispersed phase volume when droplet-fragments are formed [101]:

$$\Delta S_\gamma^{br}(d) = d^\gamma \left(N_f(d)^{\frac{3-\gamma}{3}} - 1 \right) \tag{7.46}$$

where N_f is the number of fragments formed. Since the break-up rate K_{br} is conversely analogous to the break-up time τ_{br}, the break-up source term becomes:

$$h_{br}^{(\gamma)} = \int_0^\infty \left[\frac{d^\gamma}{\tau_{br}(d)} \left(N_f\,(d)^{\frac{3-\gamma}{3}} - 1 \right) \right] n\,P(d)\,\mathrm{d}d \quad \text{for } d > d_{cr} \tag{7.47}$$

Coalescence depends not only on collisions but also on coalescence probability, thus the generic form of the coalescence source term is [15]:

$$h_{cl}^{(\gamma)} = \int_0^\infty \int_0^\infty K_{cl}(d,d')\Delta S_\gamma^{cl}(d,d')n^2 P(d)P(d')\mathrm{d}d\mathrm{d}d' \tag{7.48}$$

where K_{cl} is the collision rate, whose detailed description follows. The change in $S\gamma$ due to a single coalescence event, $\Delta S_\gamma^{cl}(d,d')$, between two droplets of diameters d and d', respectively, can be expressed as [15]:

$$\Delta S_\gamma^{cl}(d,d') = d^\gamma \left[\left(1 + \left(\frac{d'}{d} \right)^3 \right)^{\gamma/3} - \left(1 + \left(\frac{d'}{d} \right)^\gamma \right) \right] \tag{7.49}$$

or, by involving a uniform distribution with an equivalent mean diameter d_{eq}:

$$\Delta S_\gamma^{cl} = d_{eq}^\gamma (2^{\gamma/3} - 2) \tag{7.50}$$

Consequently, the coalescence source term becomes [3]:

$$h_{cl}^{(\gamma)} = (2^{\gamma/3} - 2) \left(\frac{6\,\varphi}{\pi} \right)^2 K_{cl}(d_{eq})\, d_{eq}^{\gamma-6} \tag{7.51}$$

The coalescence rate $K_{cl}(d_{eq})$ is typically given as [15]:

$$K_{cl}(d_{eq}) = K_{coll}(d_{eq})\,P_{coal}(d_{eq}) \tag{7.52}$$

where $P_{coal}(d_{eq})$ is the coalescence probability of a single collision event, which is a function of the equivalent diameter, d_{eq}, and $K_{coll}(d_{eq})$ is analogous to the product $k_{coll}\ u_{rel}(d_{eq})$, representing the Smoluchowski collision rate [3].

Finally, the coalescence source term is given as:

$$h_{cl}^{(\gamma)} = (2^{\gamma/3} - 2) \left(\frac{6\,\varphi}{\pi} \right)^2 k_{coll}\ u_{rel}(d_{eq})\ P_{coal}(d_{eq})\ d_{eq}^{\gamma-4} \tag{7.53}$$

7.3.3 Simulations

Our research team conducted a series of simulation experiments during last 15 years to study the above theoretical approach at laboratory scale. The simulations were carried out using a FORTRAN code especially developed for this

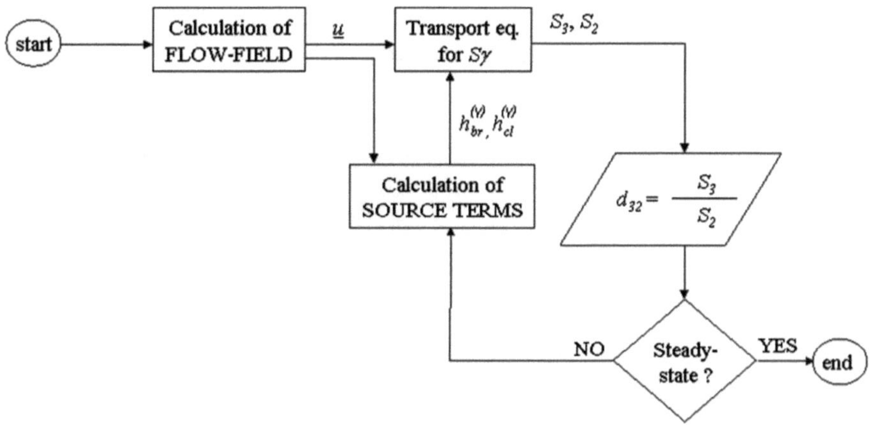

Fig. 7.13 Flowchart of the simulation algorithm

work. Calculations were performed on a simple PC equipped with an Intel© Duo Core processor and 2 GB of RAM, using Windows XP© operating system. On average, the calculations took about 12 h.

To adequately simulate the above problem, the algorithmic procedure shown in Fig. 7.13 was used. The procedure followed was: firstly the characteristics of the flow in a porous medium were calculated using standard CFD methods, and subsequently the local velocity field obtained was used in a transport equation to analyze the evolution of the domain size distribution. Although the algorithm aims at a steady-state solution, the intermediate results before reaching steady-state conditions could be used to study transient behavior of the mean droplets, by taking into account only the steady-state flow field. For the numerical solution of the transport boundary value problems, a non-uniform finite differences scheme with up winding was used for discretization in space, with the resulting linear systems of equations solved again using SOR.

To define a realistic domain for the solution of the flow and transport problem, a porous medium was constructed in the form of a spherical particle assemblage. Specifically, representation of the biphasic domains considered was achieved by the random deposition of spheres of a given radius in a box of specified dimensions as presented in Fig. 7.14, depicting a sample medium of $\varepsilon = 0.72$. The structure was digitized and the phase function was determined to obtain the specified porosity. The size of the digitized domains was $102 \times 102 \times 102$ and the length of the simulation box was approximately ten times the sphere radius.

To obtain the velocity field it is necessary to solve the Stokes equations:

$$\nabla p = \mu \nabla^2 \underline{u} \tag{7.54}$$

$$\nabla \cdot \underline{u} = 0 \tag{7.55}$$

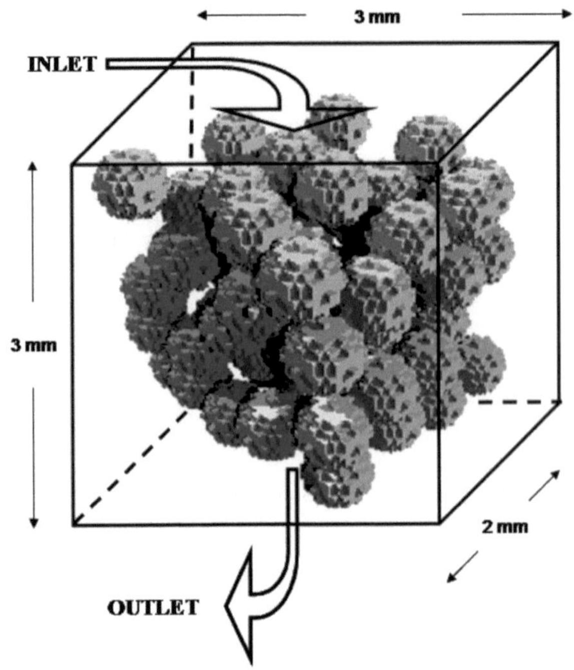

Fig. 7.14 Stochastically constructed 3-D porous medium. (Sample medium of $\varepsilon = 0.72$)

where $\underline{u}$, p, and μ are the velocity vector, pressure field and fluid viscosity, respectively. The procedure for solving the three dimensional Stokes flow problem is rather complicated in the vicinity of a porous structure [51, 54]. In the present study, a staggered marker-and-cell (MAC) mesh was involved, the pressure was defined at the cell center, and the velocity components were defined along the corresponding face boundaries. The resulting linear system of equations was solved by a successive over-relaxation (SOR) method. An initial estimate of p was determined by solving a Laplace equation. Next, the velocity vector $\underline{u}$ was calculated from the corresponding momentum balance and the continuity equation $\nabla \cdot \underline{u} = 0$. The pressure was corrected using an artificial compressibility equation of the form:

$$\frac{dp}{dt} = \beta \nabla \cdot \underline{u} \tag{7.56}$$

Essentially, the method adds an artificial density time derivative that is related to the pressure by an artificial equation of state $p = \beta \rho$, where β is an artificial compressibility factor. Similar to the compressible momentum equation, $c = \beta^{1/2}$ is an artificial speed of sound and for stability reasons during the iterative procedure, its magnitude should be such that the respective artificial Mach number, $M = \frac{R}{c} \max_D \left(\sum_i u_i^2 \right)^{1/2}$ is low ($M \ll 1$), where R is the relevant Reynolds number. In the limiting case of $R \to 0$, which is the present case, any finite value of β should meet this criterion. Thus, $\beta = 1$ was chosen here, although it has been evident that the exact value cannot have any effect on the final (steady state) results since at steady state the artificial density time derivative is zero. The above steps were repeated until convergence was reached. This numerical scheme to determine the velocity field has been widely validated in terms of both the velocity field and the corresponding permeability [2, 20].

Regarding boundary conditions, one of the six surfaces of the simulated domain was considered an inlet, while the opposite surface was assigned as an outlet. The other four domain surfaces were assumed to be spatially periodic, both in geometrical and physical (flow and transport) terms.

The differential equations for the flow-field were integrated with the following initial and boundary conditions:

$$\underline{u}(\underline{r},\, t = 0) = 0 \tag{7.57}$$

$$P(\underline{r},\, t = 0) = P_0 \tag{7.58}$$

$$u(\underline{r} = \text{void space of the inlet},\, t) = u_{in} \tag{7.59}$$

$$\nabla \cdot \underline{u}(\underline{r} = \text{void space of the inlet},\, t) = 0 \tag{7.60}$$

$$P(\underline{r} = \text{void space of the inlet},\, t) = P_{in} \tag{7.61}$$

$$P(\underline{r} = \text{void space of the outlet},\, t) = P_0 \tag{7.62}$$

$$\underline{u}(\underline{r} = \text{void/solid interfaces},\, t) = 0 \tag{7.63}$$

Initial conditions (7.57) and (7.58) indicate a quiescent mixture in the void space of the porous medium, while boundary conditions (7.59–7.61) impose a uniform (plug) flow at the inlet. Obviously, non-zero values appear only on the surfaces of the void space, since velocity and pressure equal zero at the solid phase. Boundary condition (7.62) imposes the necessary pressure gradient for the Stokes flow, while Eq. 7.63 expresses the non-slip condition at the fluid/solid interfaces throughout the porous medium.

The transport equation (7.42) was integrated with the following initial and boundary conditions:

$$S_\gamma(\underline{r},\, t = 0) = 10^{-16} \tag{7.64}$$

$$S_\gamma(\underline{r} = \text{void space of the inlet},\, t) = (S_\gamma)_{in} \tag{7.65}$$

$$\nabla \cdot S_\gamma\, (\underline{r} = \text{void space of the inlet},\, t) = 0 \tag{7.66}$$

$$S_\gamma(\underline{r} = \text{void space of the outlet},\, t) = (S_\gamma)_{out} \tag{7.67}$$

$$\nabla \cdot S_\gamma\, (\underline{r} = \text{void/solid interfaces},\, t) = 0 \tag{7.68}$$

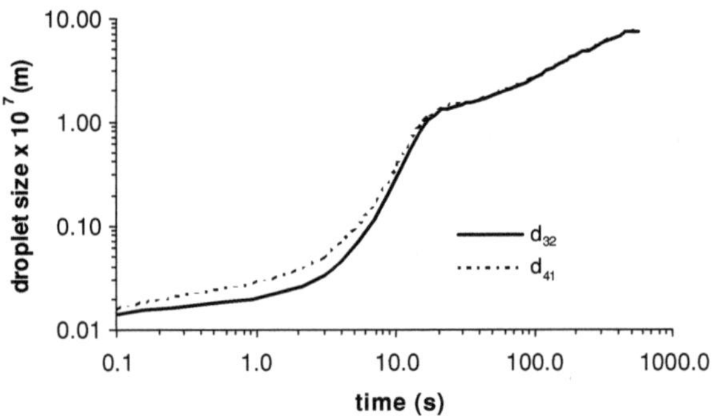

Fig. 7.15 Evolution of the average droplet size for two different sets of $S\gamma$ values

The initial condition (7.64) indicates a very low initial droplet population, while boundary conditions (7.65–7.67) impose a gradient for the $S\gamma$. Equation 7.68 expresses neutral behavior at the fluid/solid interfaces, while spatial periodicity is again assumed for the domain's other outer surfaces. Note that the above conditions are independent of the γ values.

7.3.4 Algorithm Validation and Findings

To examine the validity of the solution scheme, the algorithm was tested against its consistency. More precisely, different sets of $S\gamma$ values should lead to similar distribution parameters, i.e. to the same mean droplet size, by using Eq. 7.36. Such a test was performed by carrying out simulations for two specific sets of $S\gamma$ values, namely S_1, S_2, S_3 and S_4. These distributions lead to mean droplet size through $d_{32} = \dfrac{S_3}{S_2}$ and $d_{41} = \left(\dfrac{S_4}{S_1}\right)^{\frac{1}{3}}$, which should theoretically be the same. The mesh and time step size were the same in both cases.

Figure 7.15 depicts this comparison and the two curves represent the evolution of d_{32} and d_{41} through time. Since the results are very similar, this plot proves the self-consistency of $S\gamma$ modeling.

Since droplet size is represented by d_{32}, the solution scheme is hereafter used only for $\gamma = 2$ and $\gamma = 3$, thus the relevant source terms become:

$$h_{br}^{(2)} = \int_0^\infty \left[\frac{d^2}{\tau_{br}(d)}\left(\sqrt[3]{N_f(d)} - 1\right)\right] n\, P(d)\, \mathrm{d}d \tag{7.69}$$

$$h_{cl}^{(2)} = -0.740079 \left(\frac{6\,\varphi}{\pi}\right)^2 k_{coll}\, u_{rel}\left(d_{eq}\right) P_{coal}\left(d_{eq}\right) d_{eq} \tag{7.70}$$

$$h_{br}^{(3)} = 0 \tag{7.71}$$

$$h_{cl}^{(3)} = 0 \tag{7.72}$$

Table 7.4 Parameters used in the simulations

Parameter	Units	Expression/value	Reference/comment
$N_f(d)$	(–)	$2.69 f_1(\lambda)\left(\frac{d}{d_{cr}}\right)^3$	[32]
$\tau_{br}(d)$	(s)	$\frac{2\eta_c d}{\sigma} f_2(\lambda)$	[32]
$f_1(\lambda), f_2(\lambda)$	(–)	• 1 for the current case • Other values/expressions for the specific case studied	
d_{cr}	(m)	Eq. 7.44	Calculated at each point and time step
σ	(N m^{-1})	0.0728	Since the current work is purely theoretical, the continuous phase is supposed to be water
Ca_{cr}	(–)	1	[72]
η_c	(centipoises)	0.89	Since the current work is purely theoretical, the continuous phase is supposed to be water
$P(d)$	(–)	$\frac{1}{d\,\hat{\sigma}\sqrt{2\pi}}\exp\left(-\frac{(\ln d - \ln \overline{d})^2}{2\,\hat{\sigma}^2}\right)$	[52]
$\hat{\sigma}$	(–)	0.1	[15, 52]
$\overline{d}$	(m)	3 x 10^{-8}	[15, 52]
$k_{coll}\ u_{rel}(d_{eq})$	(m s^{-1})	$0.6667\underline{u}$	[15, 52]
$P_{coal}(d_{eq})$	(–)	Follows the log-normal distribution accordingly to the droplets' diameters	[15, 52]
d_{eq}	(m)	$\left(\frac{6\varphi}{\pi S_\gamma}\right)^{\frac{1}{3-\gamma}}$	[3]
φ	(–)	$\frac{4}{3}\pi(d_{32})^3 \frac{1}{0.03\times0.03\times0.02} n$	Calculated at each time step
n	(m^{-3})	whole algorithm and Eq. 7.33	Calculated once from the solution for S_0

For the results presented here, the simulation's parameters are presented in Table 7.4.

The time evolution of the average characteristic droplet size, d_{32}, is presented in Fig. 7.16 up to the critical time where the system reaches the steady-state situation. The maximum size is observed at $t = 550$ s, mainly due to competition between the hydrodynamic phenomena, i.e. break-up and coalescence. At $t = 900$ s, the droplet size distribution levels off taking a rather constant value for longer time scales.

This dynamic behavior of the system is clarified in Fig. 7.17 which shows the relative significance of the source terms. The individual source terms have been normalized by their maximum value, i.e. $h_{br}^{(\gamma)} = h_{br}^{(\gamma)}/MAX\left\{h_{br}^{(\gamma)}\right\}$, $h_{cl}^{(\gamma)} = h_{cl}^{(\gamma)}/MAX\left\{h_{cl}^{(\gamma)}\right\}$, to fit the same graph. However, their initial values are non-zero because the model assumes

Fig. 7.16 Characteristic average droplet size (d_{32})

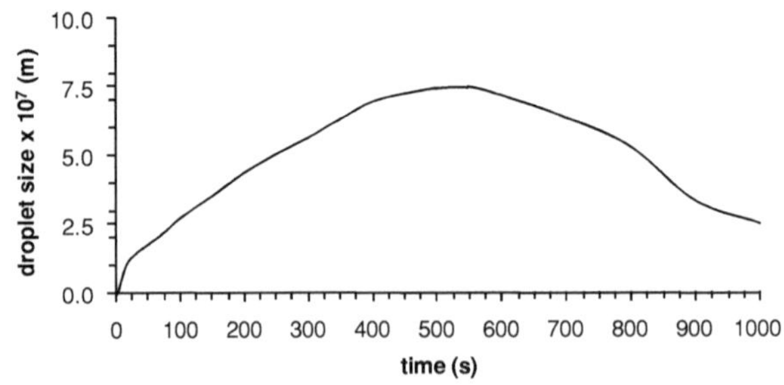

Fig. 7.17 Relative significance of source terms

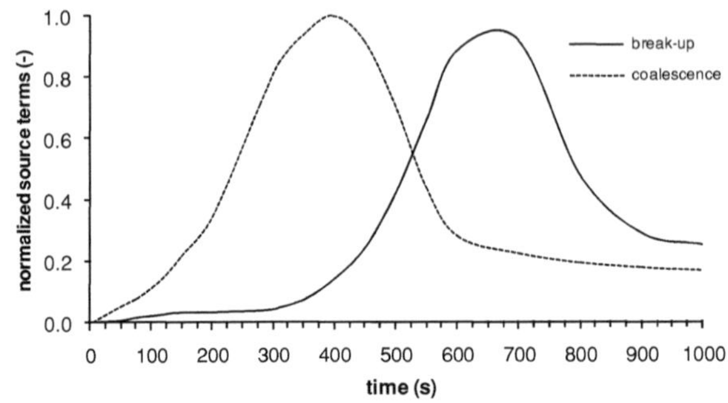

that droplet size evolution also initiates from a non-zero value (see Fig. 7.16). As shown in Fig. 7.17 during the first 550 s, coalescence dominates over break-up as the particles are quite small, therefore, droplet size increases. At $t = 550$ s the contributions of break-up and coalescence are balanced, thus particle size reaches a peak. After 550 s, break-up becomes more significant than coalescence, thus particle size decreases until $t = 900$ s. It is worth noting that the time period considered in the simulation is critical for result validity. Obviously, the simulation time is strongly dependent on the specific medium's geometry and the flow characteristics. The simulation time for the above presented results was $\mathrm{t} = 1{,}000$ s for the given porosity ($\varepsilon = 0.72$) and velocity field ($u = 1$ m/s). A more detailed study of the influence of these parameters follows.

Although porosity has not been directly included in the specified transport equations, it is a crucial parameter for the microstructure transport phenomena since it affects significantly the local flow field and mass transport regime through the void space in the porous medium. The effect of porosity is presented in Figs. 7.18 and 7.19. The time instant when droplet diameter reaches the maximum value, t_{max}, is presented in Fig. 7.18 as a function of the medium's porosity. It is notable that t_{max} varies little with porosity, and this is because droplet size is much smaller than mean pore diameter, leaving therefore the droplets unaffected by the presence of the solid phase, even for the lowest porosity value. Furthermore, lower porosity values correspond to higher coalescence probabilities because the lower

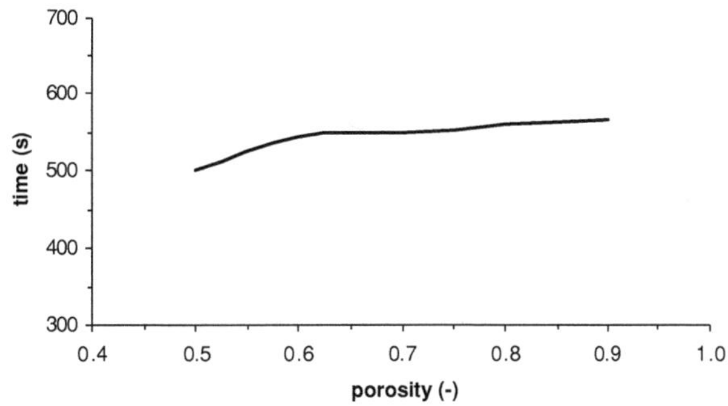

Fig. 7.18 Effect of porosity on the time when MAX{d_{32}} is observed

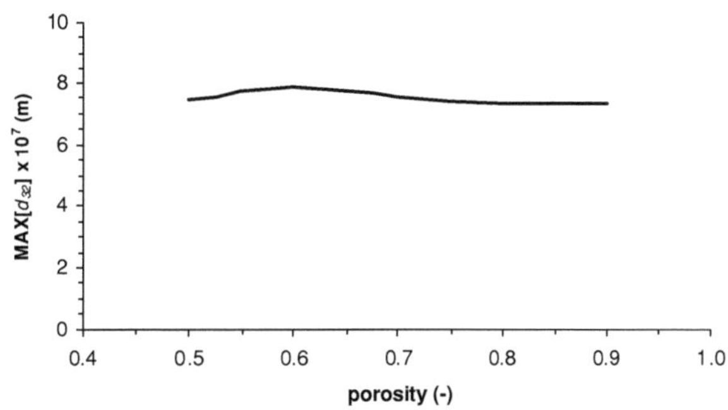

Fig. 7.19 MAX{d_{32}} as a function of porosity

the porosity, the narrower the path for a bubble to escape without any collision and, therefore, the higher the coalescence source term. In addition, break-up is not affected significantly by porosity, because $h_{br}^{(\gamma)}$ is non-zero only for diameters larger than a critical diameter value, whose order of magnitude is 10^{-7} m which corresponds to porosity values of order of magnitude 10^{-5} for feasible characteristic lengths. Actual porosity values range from 0.5 to 0.9, assuring that $h_{br}^{(\gamma)}$ is not practically affected by the porosity. Therefore, the only influence on decreasing porosity is the enhancement of $h_{br}^{(\gamma)}$ which corresponds to earlier times for maximum diameters with lower porosity. However, this influence on t_{max} does not correspond to any significant change in the absolute value of maximum diameter, as shown in Fig. 7.19. Indeed, maximum values of source terms remain rather unaffected by porosity, leading to a fairly constant maximum diameter, despite the fact that the time when these maximum values occur varies with porosity, as discussed previously.

Finally, Figs. 7.20 and 7.21 show the effect of flow intensity on droplet size. As depicted in Fig. 7.20, an increase in inlet velocity value corresponds to delayed appearances of maximum diameters. Indeed, as velocity increases i.e. convection dominates diffusion, it is easier for droplets to collide, therefore, the coalescence source term increases. Furthermore, a very low increment of the break-up source term is also expected because the higher the velocity, the easier for a droplet to

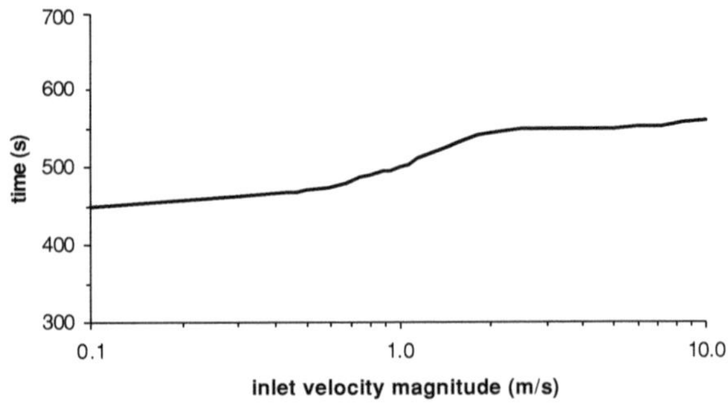

Fig. 7.20 Effect of inlet velocity magnitude on the time when MAX{d_{32}} is observed

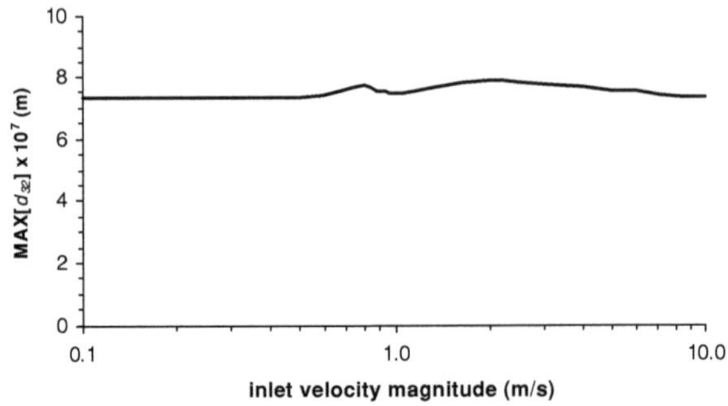

Fig. 7.21 MAX{d_{32}} as a function of inlet velocity magnitude

reach and interfere with the solid phase. As the phenomenon dominantly influenced by the flow is coalescence, maximum diameter is observed later for higher convective regimes. Again, and for the same reasons as those discussed above, this does not seem to affect significantly the maximum diameter (see Fig. 7.21).

7.4 Fluid Flows Through Porous Media in Fuel Cells

During the last decades, gradually increasing global energy demands, rising environmental and health concerns, and depletion of natural resources, have intensified the need to discover new energy production technologies. These new technologies should utilize current fuels more effectively by raising the efficiency of the energy production units, and be more environmentally friendly by reducing CO_2 emissions. Fuel cells seem to be one potential answer to this goal and offer energy policy makers energy independence under fully environmentally friendly conditions.

Fuel cells are electrochemical devices that directly convert chemical energy from gaseous fuels into electricity by overcoming Carnot limitations. Theoretically, they can reach high efficiencies and, unlike batteries, are able to provide a continuous supply of electric power when replenished with fuel. Furthermore, fuel cells are very compact units without moving parts, therefore ensuring silent operation, fewer material design constraints and reasonable portability.

Of the fuel cells available, SOFCs present considerable advantages due to their high operating temperature. These high temperatures favor electrochemical reaction kinetics permitting the use of low cost metal catalysts and they promise higher efficiencies when recovering the high energy waste heat (electrical efficiencies of 45–50% can be achieved). Furthermore, SOFC operation allows external or even internal reforming reactions to occur, thus fuels such as carbon monoxide and hydrocarbons considered poisonous to low temperature fuel cells, can be used with minimal fuel processing [1, 33, 48, 55, 76]. Despite these benefits, they do present some drawbacks regarding the materials used caused mainly by the high temperatures, such as thermal expansion of the different fuel cell components and stability issues due to mechanical deficiencies and catalyst deactivation. However, much effort has been given to fabricating new, cost effective materials with improved physical and electrochemical characteristics that would enhance both SOFC performance and stability [95].

Due to the variety of their advantages, SOFCs are considered ideal candidates for various applications. As mentioned previously, they operate at high temperatures and the waste thermal energy (waste heat) produced can be recovered in combined heat and power systems (CHP) to produce electricity and power in large-scale distributed power generation systems as well as in small-scale domestic heat and power production units, taking full advantage of the fuel used. SOFCs are also considered to be useful as auxiliary power units (APUs) for various electrical systems in cooking and transportation, such as in vehicle air conditioning and portable electronics, i.e. cell phones and personal computers. In addition, SOFCs can be used for the chemical cogeneration of electricity and chemical compounds with the use of appropriate materials. They could also be an alternative choice for remote distributed power generation either in areas with no grid supply, such as isolated islands, or when local power production is necessary, such as small power units [19, 63, 67, 84].

According to their temperature operation, SOFCs can be classified into low-, intermediate- and high-temperature types with operating temperatures between 773–823 K, 823–1,073 K and 1,073–1,273 K, respectively. Regarding geometrical aspects and design, one of the most commonly used geometries is the tubular one, initially introduced by Siemens-Westinghouse [86]. The operation principal here is that the tubular fuel cell comprises a unique tube, the air is supplied internally, and the fuel is thought to flow in the surroundings. Another design is the planar geometry in which the PEN assembly is surrounded by the collectors [43, 74]. According to the flow direction of air and fuel, planar geometry SOFCs can be subdivided into (1) co-flow planar SOFCs, where air and fuel flow in the same direction, (2) counter-flow, where they flow in opposite directions, and (3) cross-flow, where they flow perpendicular to each other [25]. Tubular and planar geometries can be both electrolyte or electrode (anode or cathode) supported [4, 92].

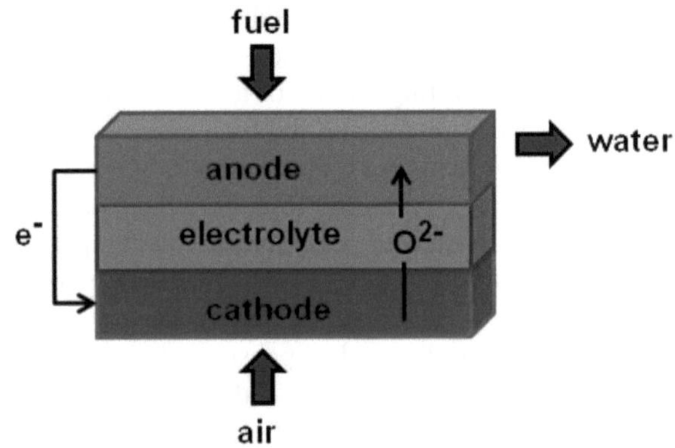

Fig. 7.22 A typical SOFC configuration

7.4.1 Solid Oxide Fuel Cell Configuration

A typical SOFC consists of an anode and a cathode compartment between which a catalyst (electrolyte) layer exists. Air flow enters one gas channel and the other channel is continuously supplied by a feeding mixture (usually a hydrogen-rich mixture containing some higher hydrocarbons with carbon monoxide and dioxide). The electrolyte has high ionic and minimum electronic conductivity and it prevents fuel crossover and mixing of the gas supplies. The fuel is electrochemically oxidized by the oxygen ions, O^{2-}, formed at the cathode electrode by the oxygen reduction reaction. These ions migrate through the electrolyte to the anode. It is evident that the same number of electrons must be transferred through the electrolyte in order to preserve equilibrium between the electrodes. The electrons released at the anode compartment flow via an external electrical circuit. In most cases, the principle operation of a typical fuel cell incorporates the formation of water from hydrogen and oxygen, and the production of electricity and heat. A typical configuration of an SOFC is presented in Fig. 7.22.

7.4.2 Electrochemical and Surface Reactions

The most significant advantage of SOFCs compared to other fuel cell types, is that they allow wide fuel flexibility due to high operational temperature and the materials used. As a result, apart from hydrogen (H_2), SOFCs can be fed with carbon monoxide (CO) and synthesis gas (H_2 and CO of various ratios), as well as natural gas and hydrocarbon fuels (coal fuels). The electrochemical reactions take place at the interface between the catalyst, the ionic conductor, and the gas phase– the so-called three-phase boundary layer (TPB). The electrochemical oxidation reactions of H_2 and CO occurring at the anode electrode can be described by:

$$H_2 + O^{2-} \rightarrow H_2O + 2e^- \tag{7.73}$$

$$CO + O^{2-} \rightarrow CO_2 + 2e^- \tag{7.74}$$

Experiments have shown that the electrochemical oxidation rate of H_2 is significantly higher (1–2.5 times) than that of CO [18, 71, 91], thus a large number of existing models neglect the contribution of CO in the production of current density.

With the use of appropriate anode cermets, such as Cu-YSZ and Ni-YSZ, direct oxidation of hydrocarbons and/or alcohols is possible in SOFCs [47, 77, 107]. In general, assuming that the fuel can be expressed as $C_nH_mO_p$ where $p = 0$ or $p = 1$ denoting hydrocarbon or alcohol fuels, respectively, the electrochemical oxidation reactions can be expressed as:

$$C_nH_m + (2n + 0.5m)O^{2-} \rightarrow nCO_2 + 0.5mH_2O + (4n + m)e^-, \text{ where } p = 0 \tag{7.75}$$

$$C_nH_mO + (2n + 0.5m - 1)O^{2-} \rightarrow nCO_2 + 0.5mH_2O + (4n + m - 2)e^-, \text{ where } p = 1 \tag{7.76}$$

Atmospheric air (humidified or not) and, rarely, pure O_2 is often introduced into the cathode gas channel. At the cathode electrode, the electrochemical oxygen reduction taking place can be written as:

$$0.5O_2 + 2e^- \rightarrow O^{2-} \tag{7.77}$$

and in case of $C_nH_mO_p$ fuels the reaction is:

$$(n + 0.25m)O_2 + (4n + m)e^- \rightarrow (2n + 0.5m)O^{2-}, \text{ where } p = 0 \tag{7.78}$$

$$(n + 0.25m - 0.5)O_2 + (4n + m - 2)e^- \rightarrow (2n + 0.5m - 1)O^{2-}, \text{ where } p = 1 \tag{7.79}$$

Apart from the electrochemical reactions, in practice, numerous homogeneous or heterogeneous chemical reactions may appear in an SOFC. The most common reactions included in simulation models are presented below. In general, the water gas shift reaction (WGS):

$$CO + H_2O \leftrightarrow CO_2 + H_2 \tag{7.80}$$

is applied when both hydrogen and carbon monoxide are present, and is usually assumed to be in equilibrium.

To effectively use hydrocarbon fuels they should be partially reformed to H_2 and CO by the internal or external steam reforming reactions as follows [1]:

$$C_nH_m + nH_2O \rightarrow nCO + (0.5m + n)H_2, \text{ where } p = 0 \tag{7.81}$$

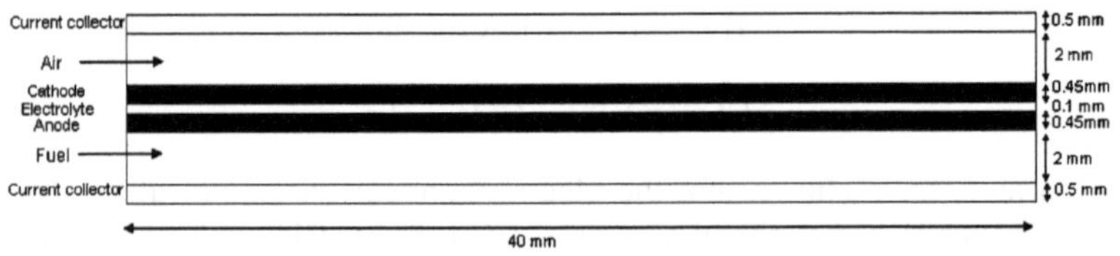

Fig. 7.23 Simulated SOFC geometry

$$C_nH_mO + (n-1)H_2O \rightarrow nCO + (0.5m + n - 1)H_2, \text{ where } p = 0 \quad (7.82)$$

Additionally, the presence of hydrocarbons may lead to carbon formation. To study the adverse effects of carbon on catalyst deactivation, the Boudouard reaction [7, 41, 66, 96, 104]

$$2CO \leftrightarrow C + CO_2 \quad (7.83)$$

and cracking reactions (where applicable)

$$C_nH_m \leftrightarrow nC + 0.5mH_2 \quad (7.84)$$

may be introduced.

7.4.3 Transport Phenomena in Gas Channels

The fundamental transport phenomena occurring in gas channels are flow and heat transfer, mass transport and charge transfer. These are described by the equations of continuity, momentum and species (neutral or charged) conservation equation. Therefore, the velocity field, temperature profile, gas composition and electric potential distribution of the fuel cell can be calculated. For better insight on the processes occurring, fuel cell geometry can be divided into two general regions. The first is the non-porous region, which refers to the gas channels, and the second includes all the porous parts of the fuel cell i.e. anode and cathode electrode, the dense electrolyte and even the current collectors. A two-dimensional cut of the fuel cell is depicted in Fig. 7.23, where the dimensions have been chosen in accordance with Ramakrishna et al. [83].

Firstly, for non-porous regions the mass conservation equation (continuity equation) can be written as:

$$\frac{\partial \rho}{\partial t} + \nabla \cdot (\rho \underline{U}) = 0 \quad (7.85)$$

where $\underline{U}$ is the velocity vector, ρ is the fluid density, and t is time.

Velocity in fuel cell gas channels is usually quite low, so the assumption of laminar flow can be considered accurate. As a result, for compressible fluids and neglecting gravitational effects, the momentum equation is:

$$\frac{\partial}{\partial t}(\rho \underline{U}) + \nabla \cdot (\rho \underline{U}\,\underline{U}) = -\nabla P + \nabla \cdot \overline{\tau} \tag{7.86}$$

where P is the pressure, and $\overline{\tau}$ is the shear stress tensor.

By neglecting radiation, the energy conservation equation becomes:

$$\frac{\partial}{\partial t}(\rho h) + \nabla \cdot (\rho \underline{U} h) = \nabla \cdot (k \nabla T) + \frac{\partial P}{\partial t} + \dot{Q} \tag{7.87}$$

where h is the total enthalpy of the fluid, k is the thermal conductivity of the mixture, and $\dot{Q}$ describes the additional thermal sources due to exothermic/endothermic reactions. The mass fraction of each gas species can be computed by the species conservation equation as follows:

$$\frac{\partial}{\partial t}(\rho Y_i) + \nabla \cdot (\rho \underline{U} Y_i) = \nabla \cdot \underline{J_i} + \dot{\omega}_i \tag{7.88}$$

where Y_i are the mass fractions of the ith chemical species, $\dot{\omega}$ is the production/destruction rate of the i – th chemical species in gas phase due to homogeneous reactions, and $\underline{J_i}$ is the species i mass diffusion flux.

Several approaches are used in the literature to calculate the mass diffusion flux, but this is beyond the scope of this analysis. The simplest model used to calculate the species diffusion flux is Fick's law:

$$\underline{J_i} = \rho D_i \nabla Y_i \tag{7.89}$$

where D_i is the mass diffusion coefficient of species i in the mixture.

Although several models for the estimation of diffusivity have been presented in the literature, in the present study the mass diffusion coefficient for gas channels is obtained through the Schmidt number, Sc, according to the relation:

$$D_i = \frac{\mu}{Sc} \tag{7.90}$$

where μ is the viscosity of the gas species.

The above differential equations are all strongly coupled and should be integrated along with the appropriate boundary conditions, which are application dependant.

7.4.4 Transport Phenomena in Porous Media

Porous materials are widely used in SOFC applications and particularly at the anode and cathode electrodes, since they present large catalytic surface areas. All the transport phenomena described above should be modified slightly to incorporate the porous nature of the electrodes and simulate the transport processes since porous material structure implies mass transport limitations.

Analytically, the mass conservation equation in a porous medium can be written as:

$$\frac{\partial}{\partial t}(\varepsilon\rho) + \nabla \cdot (\varepsilon\rho\underline{U}) = 0 \tag{7.91}$$

where ε is the porosity of the medium representing the volume occupied by the pores to the total volume of the porous media.

In porous regions, the momentum conservation equation becomes:

$$\frac{\partial}{\partial t}(\varepsilon\rho\underline{U}) + \nabla \cdot (\varepsilon\rho\underline{U}\underline{U}) = -\varepsilon\nabla P + \nabla \cdot (\varepsilon\bar{\tau}) + \frac{\varepsilon^2\mu}{\kappa}\underline{U} \tag{7.92}$$

where μ is the viscosity of the fluid, and κ is the permeability representing the square of the volume to surface area ratio of the porous material. Note that the last term of Eq. 7.92 represents Darcy's law and describes the superficial velocity in the porous medium [23].

The energy conservation equation inside a porous medium, written in terms of enthalpy, takes into account thermal convection and conduction, species diffusion contributions, and chemical reaction (neutral or electrochemical) effects on temperature. It can be written as:

$$\frac{\partial}{\partial t}(\varepsilon\rho h) + \nabla \cdot (\varepsilon\rho\underline{U}h) = \nabla \cdot \left(k_{eff}\nabla T + \sum_{i=1}^{Ng}\underline{J}_i h_i\right) + \varepsilon\bar{\tau}\nabla\underline{U} + \varepsilon\frac{dP}{dt} - j_i\eta + \frac{|\underline{i}|^2}{\sigma} \tag{7.93}$$

where k_{eff} is the effective thermal conductivity of the mixture, N_g is the total number of gas species in the system, and h_i is the enthalpy of the ith species. The temperature gradient is significantly affected inside the porous material by the electrochemical reactions and consequently both Joule heating and electrical work affect energy transfer. This effect is described by the last two terms of the energy transfer equation.

By taking into account the combination of porous and solid parts of the porous medium, the effective thermal conductivity of this medium can be defined as [35]:

$$k_{eff} = -2k_S + \frac{1}{\dfrac{\varepsilon}{2k_S + k_F} + \dfrac{1-\varepsilon}{3k_S}} \tag{7.94}$$

where k_F is the thermal conductivity of fluid parts of the porous medium, and k_S is the thermal conductivity of solid parts of the porous medium.

The conservation equation of the $i - th$ gas species is given by the relation:

$$\frac{\partial}{\partial t}(\varepsilon\rho Y_i) + \nabla \cdot (\varepsilon\rho\underline{U}Y_i) = \nabla \cdot \underline{J_i} + \dot{\omega}_i \tag{7.95}$$

where Fick's model can also be applied, given as:

$$\underline{J_i} = \rho D_{i,eff} \nabla Y_i \tag{7.96}$$

where $D_{i,eff}$ is the effective mass diffusion coefficient of species i.

To consider the porosity, ε, and tortuosity, τ, of the porous medium, a number of correlations can be found in the literature such as the Daggan model. However, the most applicable is the Bruggeman correlation, defined as [79]:

$$D_{i,eff} = \varepsilon^{\tau} D_i \tag{7.97}$$

To calculate species diffusion fluxes, apart from Fick's model applied here, several other models cam be used such as the Stefan-Maxwell model for multi component systems, and the Dusty Gas model (DGM), which incorporates both the Stefan-Maxwell formulation and the Knudsen diffusion [11, 94].

Regarding charge conservation in conducting materials, the sum of all current flows should be zero based on electro-neutrality. Thus, the current conservation equation is:

$$\nabla \cdot \underline{i} = 0 \tag{7.98}$$

where $\underline{i}$ is the current density vector

However, based on Newman and Tabias [75], charge transport consists of electronic and ionic phase transports. During electrochemical reactions electrons are either transferred from the pores (ionic phase) to the solid region (electronic phase) or vice versa, i.e. electron transfer is expressed as the transfer current, j_T, where:

$$-\nabla \cdot \underline{i_F} = \nabla \cdot \underline{i_S} = j_T \tag{7.99}$$

where $\underline{i_F}$ is the current density vector flowing through the pores (ionic phase), and $\underline{i_S}$ is the current density vector flowing through the solid parts of the porous medium (electronic phase). By applying Ohm's law, the transfer current yields:

$$\nabla \cdot (\sigma_F \nabla \varphi_F) = -\nabla \cdot (\sigma_S \nabla \varphi_S) = j_T \tag{7.100}$$

where σ_F and σ_S are the ionic phase conductivity and the solid phase conductivity, respectively, and φ_F and φ_S are the ionic potential of the fluid and the electric potential of the solid, respectively.

For the non-conducting electrolyte Eq. 7.100 becomes:

$$\nabla \cdot (\sigma_F \nabla \varphi_F) + \nabla \cdot (\sigma_S \nabla \varphi_S) = 0 \tag{7.101}$$

and the current transfer j_T can be defined through the Butler-Volmer equation as:

$$j_T = \frac{j_0\left(\frac{A}{V}\right)}{\prod_{i=1}^{N}\left[C_{i,ref}\right]^{a_{ie}}}\left[\exp\left(\frac{a_a F}{RT}\eta\right) - \exp\left(-\frac{a_c F}{RT}\eta\right)\right]\prod_{i=1}^{N}\left[C_i\right]^{a_{ie}} \tag{7.102}$$

where $i._e$ indexes the electrochemical reactions, j_0 is the exchange current density, a_a and a_c are the anodic and cathodic charge transfer coefficients as determined by the Tafel plots respectively, N is the total number of reacting species, $[C_i]$ is the near wall molar concentration of the ith reacting species or expressed in mass fractions as $[C_i] = \rho \frac{Y_i}{M_i}$, where M_i is the molar weight of the $i - th$ species, $[C_{i,ref}]$ is the molar concentration at a reference state at the inlet, and a_k is the concentration exponent. The overpotential, η, can be expressed as the potential difference between the solid phase and the porous phase potential:

$$\eta = \varphi_S - \varphi_F \tag{7.103}$$

In surface reactions occurring at porous/catalyst interfaces, the volumetric reaction rate (production/destruction of a gas species i) can be computed assuming a balance between the reaction flux and the diffusion flux on the surface [50]. This rate is given as:

$$\dot{\omega} = \rho D_i \frac{Y_i - Y_{P,i}}{\delta} \left[\frac{A}{V} \right]_{eff} \tag{7.104}$$

where $Y_{P,i}$ is the mass fraction of the ith species in the pore fluid, $[A/V]_{eff}$ denotes the effective surface to volume ratio of the catalyst and represents the catalyst load, and δ is the diffusion length scale. For electrochemical reactions, the production/ destruction rate is expressed through the current transfer by the relation:

$$\dot{\omega} = (b_{PR} - b_R) \frac{j_T}{F} \tag{7.105}$$

where b_{PR} and b_R are the stoichiometric coefficients of the products and reactants, respectively.

7.4.5 Simulations

Planar SOFC geometry was used for the three-dimensional approach. This geometry consists of seven separated volumes: the fuel and air channel where the corresponding mixtures are introduced, the porous anode and cathode electrodes where the reactions occur, the dense electrolyte through which oxygen ions migrate to reach the anode electrode, and finally the anode and cathode contacts. In the present study, both H_2 and CO were assumed to react with oxygen ions, in accordance with Eqs. 7.73 and 7.74. Regarding the Butler-Volmer equation (Eq. 7.102), the exchange current density was set to $j_0 = 10^{11} \mathrm{Am}^{-3}$ for the anode and $j_0 = 10^{10} \mathrm{Am}^{-3}$ for the cathode, where the charge transfer coefficients for both electrodes were $a_a = 0.7$ and $a_c = 0.7$. The electrochemical reactions were assumed to occur only in the porous electrodes and precisely on the three-phase boundary layer. It was also assumed that no other reactions occurred either on the porous interface or in the bulk phase. The feeding stream was assumed to be a reformate mixture of 9.6% H_2, 26.0% CO, 2.16% CO_2 and 42.8% H_2O on w.t.% basis, and typical atmospheric air composition was applied at the cathode channel (23.3% O_2 and 76.7% N_2 on w.t. basis).

Regarding the boundary conditions, the mass flow rates for the anodic and cathodic mixtures were assumed to be constant and equal to 3.0×10^{-7} kg/s and 4.0×10^{-6} kg/s, respectively, while pressure of 1 atm was set at the inlets and outlets. Considering no accumulation, zero mass flux was set at the walls and outlets. Additionally, preheated mixtures of 1,173 K entered the fuel cell, while zero heat flux was set at all the other boundaries. Finally, the constant value of zero overpotential was set to the anode contact, while the overpotential of the cathode contact was set to -0.7 V.

Both electrodes were modeled as isotropic porous media and their physical characteristics are as follows [58, 83]: porosity was $\varepsilon = 0.4$ for the anode, $\varepsilon = 0.01$ for the electrolyte, and $\varepsilon = 0.5$ for the cathode; permeability was $\kappa = 10^{-12}$ $m^2$for the anode and cathode, and $\kappa = 10^{-18}$ m^2 for the electrolyte; solid phase electrical conductivity was $\sigma_S = 10,000\ \Omega m^{-1}$ for the anode, $\sigma_S = 10^{-20}\,\Omega m^{-1}$ for the electrolyte, and $\sigma_S = 7,700\ \Omega m^{-1}$ for the cathode; ionic phase electrical conductivity was $\sigma_F = 10\,\Omega m^{-1}$ for both the electrodes and the electrolyte.

To calculate the density, ρ, it was assumed that the gases followed the ideal gas law, and the kinetic theory of gases was used to estimate viscosity, μ. Specific heat, c_p, was derived by fittings to experimental JANNAF curves, and mass diffusivity was calculated by imposing 0.7 to the Schmidt number.

To achieve residual values less than 10^{-4} for all the quantities and assuming steady state conditions, the numerical solution of the transport problems was obtained by the commercial package CFD-ACE$^+$ by ESI-Group©, which is based on the finite volume method, together with all the appropriate boundary conditions. The three-dimensional fuel cell was discretized in space using a structured grid of 33, 516 cells.

7.4.6 Results and Discussion

The typical distributions for the main physical quantities (velocity, temperature mass fractions and overpotential) obtained for constant gas inlet mixture and temperature for both fuel and air mixtures are presented in Figs. 7.24, 7.25, 7.26, 7.27, 7.28, 7.29. These are contour plots in a two-dimensional cut in the middle plane of the fuel cell. Figure 7.24 shows the developed velocity profile, which was parabolic for both flow channels, satisfying the non-slip conditions applied on the walls. Due to the mass flow rates imposed, high velocity values can be seen in the air channel and lower velocity values in the fuel channel, while zero velocities are observed in the porous media. The highest value observed was approximately 5 m/s, which can be considered quite high, but necessary to satisfy the energy demands of the system in terms of cooling, and maintain active chemical reactions.

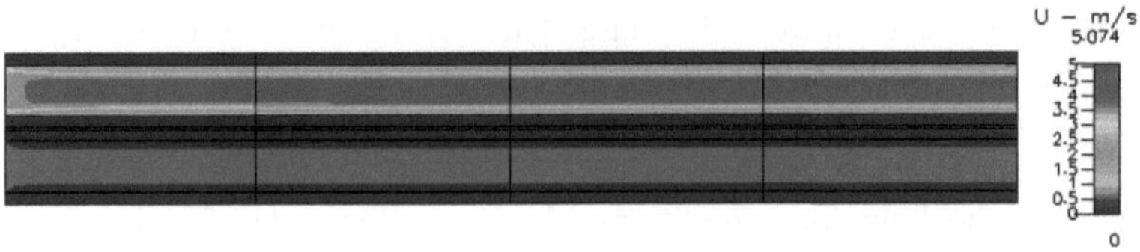

Fig. 7.24 Velocity profile

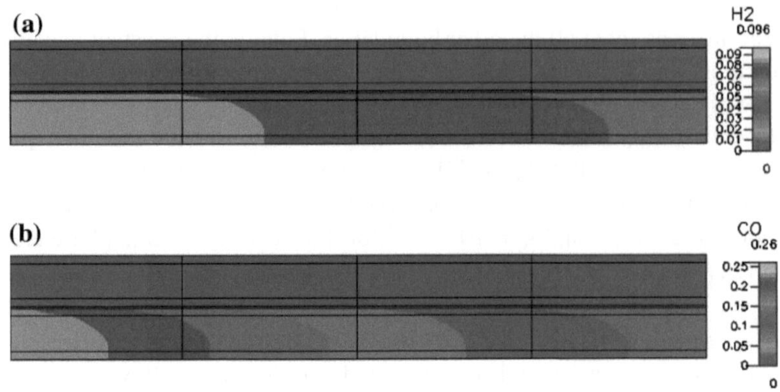

Fig. 7.25 Mass fraction profiles for (a) hydrogen and (b) carbon monoxide

The molar fractions of the major reactants and products along the fuel cell are presented in Figs. 7.25, 7.26. Figure 7.25 shows that hydrogen and carbon monoxide are consumed at the anode channel and anode electrode. Furthermore, their fractions decrease with distance, due to the electrochemical reaction controlling the system. The consumption of both species is accompanied by relative production of both CO_2 and H_2O, in accordance with the relative reaction rates.

A small depletion of oxygen can be observed at the cathode and the cathode channel (Fig. 7.27). This can be attributed to its participation in the electrochemical reactions. The O_2 gradient is fairly small since the air mass flow rate is high enough to avoid oxygen depletion that would lower fuel cell performance, while the highly convective regime does not favor the electrochemical reactions.

Temperature varies along the fuel cell length from 1,173 K imposed at the inlet boundary to 1,256 K at the outlet (Fig. 7.28). This temperature variation is strongly affected by the electrochemical reactions occurring at the anode electrode. Hydrogen and carbon monoxide electro oxidation reaction is exothermic, thus a certain amount of heat is released in the cell and transferred locally due to convection and conduction. Consequently, temperature increase can be observed along the length of the cell and it is evident that convectional effects prevail at the air channel due to the developing velocities. Finally, the over potential profile is depicted in Fig. 7.29. The over potential at the anode electrode is higher compared to that at the cathode electrode due to fuel depletion at the anode channel. It should

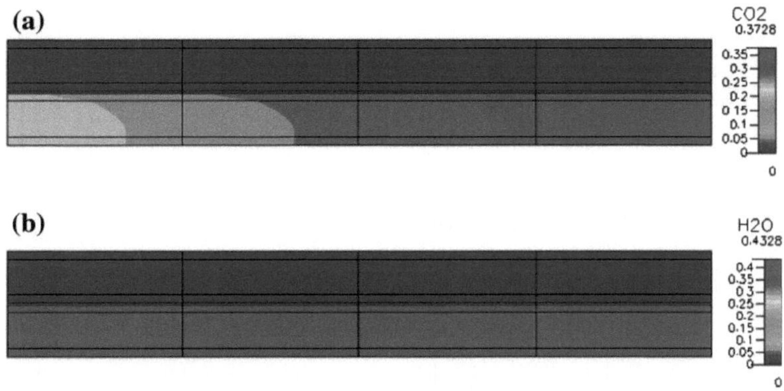

Fig. 7.26 Mass fraction profiles for (**a**) carbon dioxide and (**b**) steam

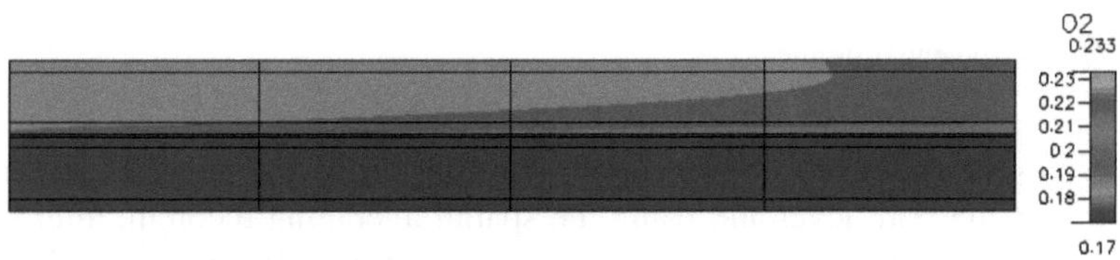

Fig. 7.27 Mass fraction profile for oxygen

also be mentioned that over potential increases along the anode electrode and this rise is followed by the reduction of active species mass fractions i.e. H_2 and CO.

7.5 Multi Phase Transport in Porous Media

Numerous physicochemical processes exist where an aqueous phase coexists and/or interacts with a non-aqueous liquid phase in a porous medium. Many of these processes have significant importance in terms of industrial and technological applications, e.g. tracer transport in petroleum reservoirs, chemical contamination of soils and aquifers, long-term interaction between liquefied or chilled foods with packaging materials. The description of such processes often relies on mathematical models as experimentation can be expensive or difficult.

In the presence of porous media, the need for a realistic description of the structure of a porous medium significantly increases the mathematical complexity of a model. However, elements of the microstructure must be captured when moving from the pore level to the macroscopic level, where process performance must often be studied. A number of techniques for up scaling from the pore-scale to the macroscopic scale in porous media have been developed. Of specific interest here is volume-averaging [14, 78, 105, 106], which is very useful when pore scales

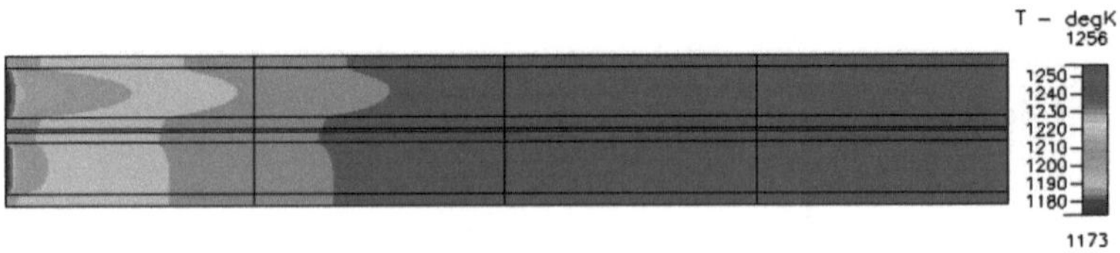

Fig. 7.28 Spatial distribution for temperature

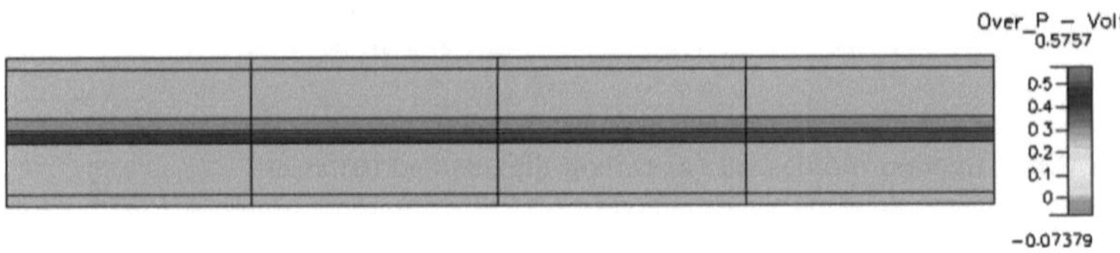

Fig. 7.29 Overpotential profile

and macroscopic scales are separated. Starting from the relevant differential equations at the pore level and using the spatial averaging theorem, after several mathematical manipulations based on scale separation [102, 103], one is lead to unit-cell problems to estimate macroscopic quantities [80, 81]. Spatial averaging techniques allow the integral effect of the influence of pore geometry on transport to be captured in the mathematical formulation.

Typically, macroscopic models involve an exchange term to describe the rate of mass transfer between the two phases. The majority of existing theoretical works derived such models based on the assumption of the mass exchange equilibrium between the aqueous and non-aqueous liquid phases [27, 28, 45, 62]. This assumption of infinitely fast diffusion in the non-aqueous phase, simplifies mathematical modeling and consequent simulations [37, 82, 88]. It is necessary to determine the validity of this assumption. When the assumption is relaxed, the resulting unit-cell problem includes mass transport between the two phases (the "aqueous" and the "non-aqueous" liquids), thus accounting for inter-phase diffusion and partitioning. Its solution provides expressions required for the macroscopic mass-transfer coefficient. Various model pore geometries are used to obtain representative results. An ab initio calculation is also presented, in which the Stokes equations are solved in typical pore geometries and compared with the volume-averaging results.

7.5.1 Theoretical Background

Consider a multi-phase domain consisting of a flowing aqueous phase (β-phase), an immobile non-aqueous liquid phase (γ-phase) and a solid phase (σ-phase), as schematically depicted in Fig. 7.30.

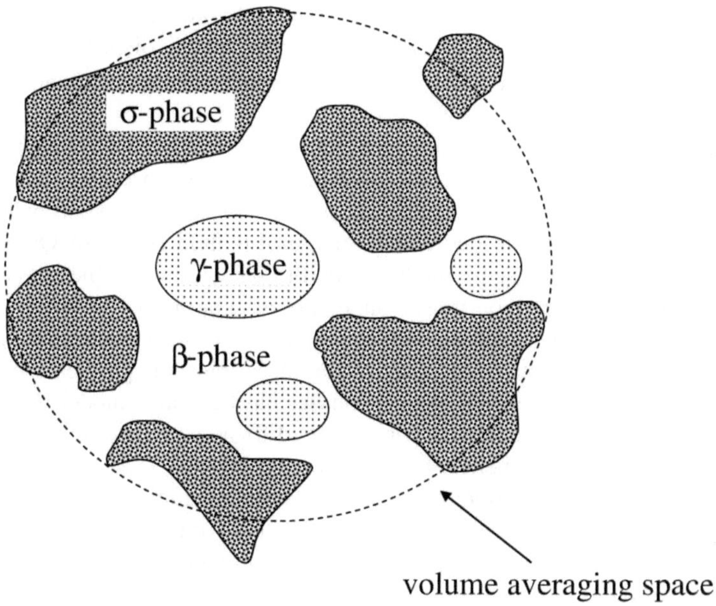

Fig. 7.30 Schematic depiction of a typical representative volume

The assumption of a motionless γ-phase is a good approximation for several applications, such as water or gas tracer transport in hydrocarbon reservoirs where the oil phase rests practically immobile (e.g. residual saturation), slow dissolution of NonAqueous Phase Liquids (NAPLs) in underground porous formations, and packing porous pellets with a wetting fluid. To relax this assumption, one should backtrack to techniques (e.g. Lattice-Boltzmann) capable of solving the flow problem while simultaneously determining the dynamic distribution of fluid phases in the considered porous domain. This approach, however, requires significant computational resources [10]. A partitioning tracer is advected by the flowing β-phase and partitioned in the immobile phase into which it is diffusing. It is assumed that the solid phase is physicochemically neutral, i.e. the tracer is neither adsorbed nor reacts with the σ-phase. The governing processes in the β-phase are diffusion and advection, and diffusion in the γ-phase. The mass exchange at the $\beta\gamma$ interface is characterized by the diffusion and partitioning properties of the tracer. The macroscopic modeling of the processes is typically described by the following advection–dispersion-reaction equations:

$$\varepsilon_\beta \frac{\partial \langle C_\beta \rangle}{\partial t} + \langle \mathbf{v} \rangle \cdot \nabla \langle C_\beta \rangle = \varepsilon_\beta D_\beta^* \nabla^2 \langle C_\beta \rangle - \alpha \left(\langle C_\beta \rangle - \frac{1}{K} \langle C_\gamma \rangle \right) \tag{7.106}$$

$$\varepsilon_\gamma \frac{\partial \langle C_\gamma \rangle}{\partial t} = \varepsilon_\gamma D_\gamma^* \nabla^2 \langle C_\gamma \rangle + \alpha \left(\langle C_\beta \rangle - \frac{1}{K} \langle C_\gamma \rangle \right) \tag{7.107}$$

where ε_β, ε_γ denote the volume fraction of the β- and γ-phase, respectively, v is the velocity vector in the β-phase, D^*_β, D^*_γ are the macroscopic dispersion coefficients in the β- and γ- phase, respectively, and K is the partitioning coefficient. Brackets denote spatial averages.

The knowledge of several parameters (e.g., of the mass-transfer coefficient and the dispersion tensor) is required to solve the above macroscopic equations. As previously noted, volume averaging can provide these expressions when scale separation exists. The main theoretical work in this field is that of Quintard and Whitaker [82], which presents methods for the calculation of the mass-transfer coefficient and the dispersion tensor under the assumption of infinitely fast diffusion in the γ-phase. This corresponds to an almost constant concentration profile in the γ-phase and permits decoupling of the mass transport problems in the two phases. Mass exchange between the two phases has been also modeled by various researchers, considering either a controlling diffusive process macroscopically described by first order kinetics, or an advection process [5, 9, 22, 38, 65, 96]. What is proposed here is a more rigorous extension that considers non-equilibrium partitioning of the tracer and mass transfer in both phases. Three-dimensional model pore geometries were considered in the following numerical illustrations, where the immobile γ-phase may be distributed as either wetting films or non-wetting aggregates (blobs). The objective is to estimate the mass transport coefficient and investigate the effect of different structural and physicochemical parameters under non-equilibrium partitioning. Otherwise, the volume averaging approach is conventional and consists of the algorithm presented in Sect. 5.5.

7.5.2 Formulation of the Problem

The pore-level transport of the tracer in the β-phase is described by the diffusion–advection equation

$$\frac{\partial C_\beta}{\partial t} + \nabla \cdot (vC_\beta) = D_\beta \nabla^2 C_\beta \tag{7.108}$$

where C_β is concentration, t is time, v is the fluid velocity, and D_β is the diffusivity in the β-phase. Since the γ-phase is assumed to be immobile, the diffusion equation describes the transport of the tracer in that phase:

$$\frac{\partial C_\gamma}{\partial t} = D_\gamma \nabla^2 C_\gamma \tag{7.109}$$

where C_γ and D_γ are concentration and diffusivity in the γ-phase, respectively. Zero-flux boundary conditions apply to the solid–liquid interfaces:

$$\mathbf{n}_{\beta\sigma} \cdot \nabla C_\beta = 0 \quad \text{at} \quad A_{\beta\sigma} \tag{7.110}$$

$$\mathbf{n}_{\gamma\sigma} \cdot \nabla C_\gamma = 0 \quad \text{at} \quad A_{\gamma\sigma} \tag{7.111}$$

as the transported species do not absorb or react at the solid. At interface $A_{\beta\gamma}$ between trapped and flowing phases, the following conditions apply:

$$C_\gamma = KC_\beta \tag{7.112}$$

$$D_\beta \mathbf{n}_{\beta\gamma} \cdot \nabla C_\beta = D_\gamma \mathbf{n}_{\beta\gamma} \cdot \nabla C_\gamma \tag{7.113}$$

where K is the partitioning coefficient. Equation 7.112 describes the partition equilibrium, while (7.113) expresses the flux continuity at the interface. This mass exchange description is more general than in previous works [82] and does not assume fast diffusion in the γ-phase as used by Quintard and Whitaker [82], De Smedt and Wierenga [88] and Gvirtzam et al. [37].

Following the volume averaging procedure [82], local concentrations and velocities are next decomposed into interstitial averages and fluctuations:

$$C_\beta = \langle C_\beta \rangle^\beta + C'_\beta \tag{7.114}$$

$$C_\gamma = \langle C_\gamma \rangle^\gamma + C'_\gamma \tag{7.115}$$

$$\mathbf{v}_\beta = \langle \mathbf{v}_\beta \rangle^\beta + \mathbf{v}'_\beta \tag{7.116}$$

which are subsequently substituted into the governing differential equations. Invoking separation of scales to discard small terms, linearizing and following Quintard and Whitaker [82], the representation below can be easily obtained when assuming an isotropic medium of uniform porosity and constant volume fractions:

$$C'_\beta = \mathbf{b}_\beta \cdot \nabla \langle C_\beta \rangle^\beta + s_\beta \left(\frac{1}{K} \langle C_\gamma \rangle^\gamma - \langle C_\beta \rangle^\beta \right) \tag{7.117}$$

$$C'_\gamma = \mathbf{b}_\gamma \cdot \nabla \langle C_\gamma \rangle^\gamma + s_\gamma \left(\frac{1}{K} \langle C_\gamma \rangle^\gamma - \langle C_\beta \rangle^\beta \right) \tag{7.118}$$

where b_β, b_γ, s_β and s_γ are closure variables satisfying specific boundary value problems.

7.5.3 The Closure Problems

It can be readily shown that variable s_β of the β-phase satisfies the boundary value problem [82]:

$$\mathbf{v}_\beta \cdot \nabla s_\beta = D_\beta \nabla^2 s_\beta - \varepsilon_\beta^{-1} \alpha \tag{7.119}$$

with boundary conditions:

$$s_\beta = 1 + \frac{s_\gamma}{K} \text{ at } A_{\beta\gamma} \tag{7.120}$$

$$D_\beta \mathbf{n}_{\beta\gamma} \cdot \nabla s_\beta = D_\gamma \mathbf{n}_{\beta\gamma} \cdot \nabla s_\gamma \text{ at } A_{\beta\gamma} \tag{7.121}$$

$$\mathbf{n}_{\beta\sigma} \cdot \nabla s_\beta = 0 \text{ at } A_{\beta\sigma} \tag{7.122}$$

and the compatibility condition:

$$\langle s_\beta \rangle = 0 \tag{7.123}$$

Similarly, variable s_γ of the γ-phase satisfies the following problem:

$$D_\gamma \nabla^2 s_\gamma = -\varepsilon_\gamma^{-1} \alpha \tag{7.124}$$

$$\mathbf{n}_{\gamma\sigma} \cdot \nabla s_\gamma = 0 \text{ at } \quad A_{\gamma\sigma} \tag{7.125}$$

$$\langle s_\gamma \rangle = 0 \tag{7.126}$$

where the mass transfer coefficient, α, is given by:

$$\alpha = \frac{D_\beta}{V} \int\limits_{A_{\beta\gamma}} \mathbf{n}_{\beta\gamma} \cdot \nabla s_\beta dA \tag{7.127}$$

The above equation can be simplified by introducing the transformation $s_\beta = 1 + \alpha\psi_{\beta s}$ and $s_\gamma = \alpha\psi_{\gamma s}$, and in dimensionless form:

$$s_\beta = 1 + \alpha^* \zeta_\beta \tag{7.128}$$

$$s_\gamma = \alpha^* \zeta_\gamma \tag{7.129}$$

where the dimensionless mass transfer coefficient is defined as $\alpha^* = \dfrac{a l_\beta^2}{D_\beta}$, and l_β is a characteristic length of the β-phase. Then, in dimensionless notation, the boundary value problems read as follows: In the β-phase:

$$Pe_\beta v \cdot \nabla \zeta_\beta = \nabla^2 \zeta_\beta - \varepsilon_\beta^{-1} \tag{7.130}$$

$$\mathbf{n}_{\beta\sigma} \cdot \nabla \zeta_\beta = 0 \quad \text{at} \quad A_{\beta\sigma} \tag{7.131}$$

where $Pe_\beta = \dfrac{\langle \mathbf{v} \rangle^\beta l_\beta}{D_\beta}$. In the γ-phase:

$$0 = \delta \nabla^2 \zeta_\gamma + \varepsilon_\gamma^{-1} \tag{7.132}$$

$$\mathbf{n}_{\gamma\sigma} \cdot \nabla \zeta_\gamma = 0 \quad \text{at} \quad A_{\gamma\sigma} \tag{7.133}$$

$$\langle \zeta_\gamma \rangle = 0 \tag{7.134}$$

where $\delta = \frac{D_\gamma}{D_\beta}$. The two problems are coupled at their interface through the condition:

$$\zeta_\beta = \frac{1}{K}\zeta_\gamma \quad \text{at } A_{\beta\gamma} \tag{7.135}$$

In this notation, the mass transfer coefficient becomes:

$$\alpha^* = -\frac{\varepsilon_\beta}{\langle \zeta_\beta \rangle} \tag{7.136}$$

One final substitution will allow for additional insight. Define:

$$\phi_\beta = \delta K \varepsilon_\gamma \zeta_\beta \tag{7.137}$$

and

$$\phi_\gamma = \delta \varepsilon_\gamma \zeta_\gamma \tag{7.138}$$

Then, the two boundary value problems take the canonical form:

$$\mathbf{Pe}_\beta \mathbf{u} \cdot \nabla \phi_\beta = \nabla^2 \phi_\beta - \Lambda \text{ in the} \beta\text{−phase} \tag{7.139}$$

$$\mathbf{n}_{\beta\sigma} \cdot \nabla \phi_\beta = 0 \quad \text{at } A_{\beta\sigma} \tag{7.140}$$

where $\Lambda = \frac{\delta K \varepsilon_\gamma}{\varepsilon_\beta}$ and

$$0 = \nabla^2 \phi_\gamma + 1 \text{ in the } \gamma\text{−phase} \tag{7.141}$$

$$\mathbf{n}_{\gamma\sigma} \cdot \nabla \phi_\gamma = 0 \quad \text{at } A_{\gamma\sigma} \tag{7.142}$$

$$\phi_\beta = \phi_\gamma \quad \text{at } A_{\beta\gamma} \tag{7.143}$$

$$\langle \phi_\gamma \rangle = 0 \tag{7.144}$$

The mass-transfer coefficient is simply:

$$\alpha^* = -\frac{\delta K \varepsilon_\beta \varepsilon_\gamma}{\langle \phi_\beta \rangle} \tag{7.145}$$

7.5.4 Results and Discussion

The above scaling relations are checked against analytical solutions and numerical simulations below. Two different unit cell geometries are used: the first involving two parallel phases of infinite extent, and the second involving a unit cube with a disordered phase distribution.

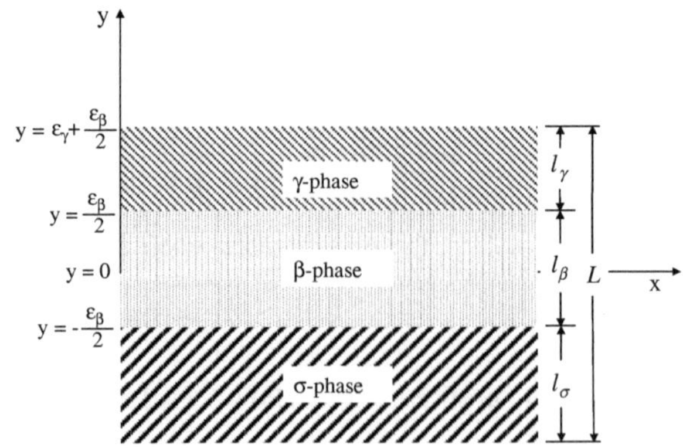

Fig. 7.31 A two-dimensional unit-cell for parallel phases

In the numerical simulations, the velocity field was computed numerically by solving the Stokes equations:

$$\nabla p = \mu \nabla^2 \mathbf{v} \tag{7.146}$$

$$\nabla \cdot \mathbf{v} = 0 \tag{7.147}$$

$$\mathbf{v} = \mathbf{0} \quad \text{at}\, A_{\beta\sigma} \tag{7.148}$$

where v, p, and μ are the velocity vector, pressure field and fluid viscosity, respectively. The procedure for solving the three-dimensional Stokes flow problem involves discretization in terms of cubic elements and was presented in Sect. 3.5. To numerically solve the transport boundary value problems, a non-uniform finite differences scheme with upwinding was used for discretization, with the resulting linear systems of equations solved again using Successive Over-Relaxation method (SOR).

The solution for simplified geometry of parallel phases assuming one-dimensional flow is presented first. In addition, it is considered that diffusion is significant only in the direction perpendicular to flow. These assumptions correspond to a Pe-independent Type II closure problem similar to that of Quintard and Whitaker [82] for sufficiently fast diffusion. The unit cell is shown Fig. 7.31 and was used to calculate properties of the Taylor-Aris dispersion problem for a passive tracer.

The boundary value problem can be solved analytically in this geometry, yielding the following expression for the dimensionless mass transfer coefficient:

$$\alpha^* = \frac{\varepsilon_\beta \varepsilon_\gamma}{A_1 + A_2 + A_3} \tag{7.149}$$

where the length L of the unit cell is used as characteristic length. (Note that in this notation $\alpha^* = \alpha^\wedge \left(\frac{l_\beta}{L}\right)^2$). The coefficients are, respectively:

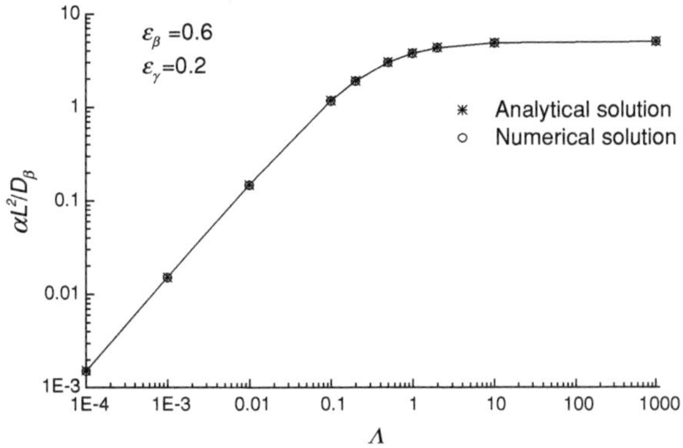

Fig. 7.32 Analytical and numerical results of the dimensionless mass transfer coefficient for parallel phases

$$A_1 = -\frac{1}{6\Lambda}\left[\left(\frac{\varepsilon_\beta}{2}+\varepsilon_\gamma\right)^3 - \frac{\varepsilon_\beta^3}{8}\right]\varepsilon_\beta^2 \tag{7.150}$$

$$A_2 = \frac{1}{2\Lambda}\left(1+\frac{\varepsilon_\beta}{2\varepsilon_\gamma}\right)\left[\left(\frac{\varepsilon_\beta}{2}+\varepsilon_\gamma\right)^2 - \frac{\varepsilon_\beta^2}{4}\right]\varepsilon_\gamma\varepsilon_\beta^2 \tag{7.151}$$

$$A_3 = \frac{\varepsilon_\beta^4\varepsilon_\gamma}{3} - \left(\varepsilon_\gamma+\frac{\varepsilon_\beta}{4}\right)\frac{\varepsilon_\beta^3\varepsilon_\gamma}{2\Lambda} \tag{7.152}$$

In the limiting case of infinitely fast diffusion in the γ-phase ($\delta \to \infty \Rightarrow \Lambda \to \infty$), the asymptotic results are $A_1 \to 0$, $A_2 \to 0$ and $A_3 \to \frac{\varepsilon_\beta^4\varepsilon_\gamma}{3}$, thus the dimensionless mass transport coefficient is $\hat{\alpha} \to \frac{3}{\varepsilon_\beta}$. This result coincides with the analytical mass transport coefficient for parallel phases found by Quintard and Whitaker [82] and shown to be independent of Peclet number.

Figure 7.32 shows the dimensionless mass transfer coefficient plotted as a function of the governing dimensionless parameter $\Lambda = \frac{\delta K \varepsilon_\gamma}{\varepsilon_\beta}$, calculated using both the analytical approach and numerical simulation. The analytical and numerical solutions are almost identical. For low values of Λ, the mass transfer coefficient increases with Λ (i.e. with increasing values of δ and/or K) in a linear fashion. At higher values, α reaches a constant value and becomes practically independent of Λ. The spatial distribution of the dimensionless scalar s_β is shown in Fig. 7.33. In the limiting case of $\Lambda \to \infty \Rightarrow \delta \to \infty$ the calculated s_β values are identical to those found by Quintard and Whitaker [82] for fast diffusion in the γ-phase with the same geometry.

Further investigation of the validity of the asymptotic scaling results in this case, is portrayed in Fig. 7.34, where the dimensionless mass transfer coefficient is plotted against Λ for various values of the fraction ε_β, with the ratio $\frac{\varepsilon_\beta}{\varepsilon_\gamma}$ kept constant.

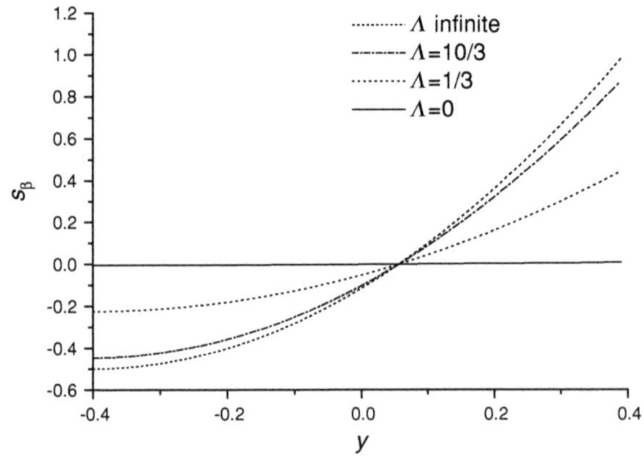

Fig. 7.33 Numerical results for s_β for parallel phases

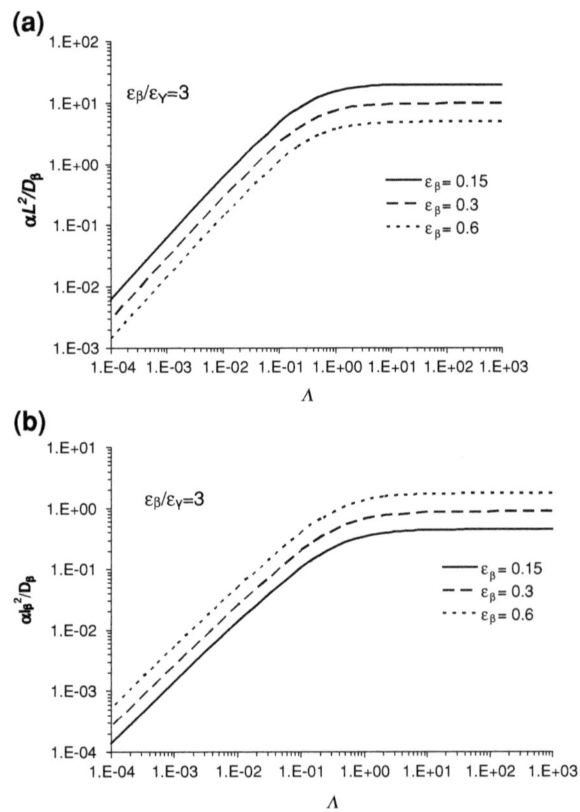

Fig. 7.34 The effect of Λ on the mass transfer coefficient, defined based either on the characteristic length L (**a**) or the characteristic length l_β (**b**) for various values of the volume fraction and for parallel phases. Static conditions

In Fig. 7.34a, the length of the cell, $L = l_\sigma + l_\beta + l_\gamma$, is used as the characteristic length. The mass transfer coefficient $\hat{\alpha}$is shown to decrease linearly with ε_β. This result is in accordance with previous investigations (e.g. see Eq. 47e in [82]) although α^* is expected to increase linearly with ε_β. The explanation lies in the

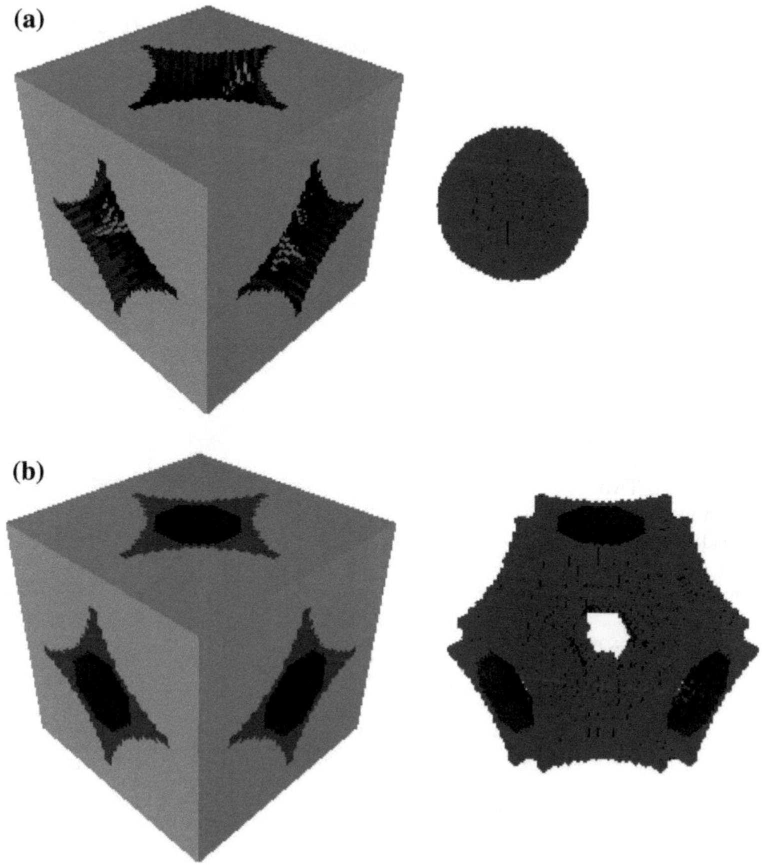

Fig. 7.35 A typical three-dimensional unit cell with the immobile fluid phase configuration in the form of (**a**) blobs, or (**b**) films. Volume fractions are $\varepsilon_\beta = 0.31$, $\varepsilon_\gamma = 0.12$. The immobile phase is also plotted separately for clarity

different characteristic lengths used to non-dimensionalize the mass transfer coefficient. It is not difficult to show that for parallel phase geometry it is $\frac{l_\beta}{L} = \varepsilon_\beta$, hence $\alpha^* = \hat{a}\varepsilon_\beta^2$. Thus, Fig. 7.34b shows the corresponding plot for α^* for various values of ε_β. Where $a^* \to 3\varepsilon_\beta$ when $\delta \to \infty \Leftrightarrow \Lambda \to \infty$, as expected.

More realistic geometries are considered in Fig. 7.35 where a typical periodic three-dimensional unit cell is shown, assumed to represent regularly packed granular porous media.

The solid grains (σ-phase) are located on the eight edges of the cell. For the immobile γ-phase, two different cases are considered:

- The immobile phase does not wet the solid, hence it is distributed in the pore space in the form of blobs.
- The immobile phase wets the solid, hence it is distributed in the pore space in the form of films.

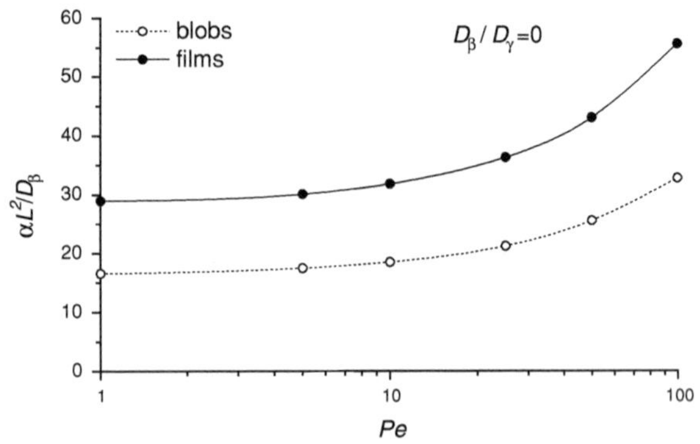

Fig. 7.36 The effect of Peclet number on the mass transfer coefficient for the limiting case of sufficiently fast diffusion in the γ-phase for the three-dimensional unit cell of Fig. 7.35

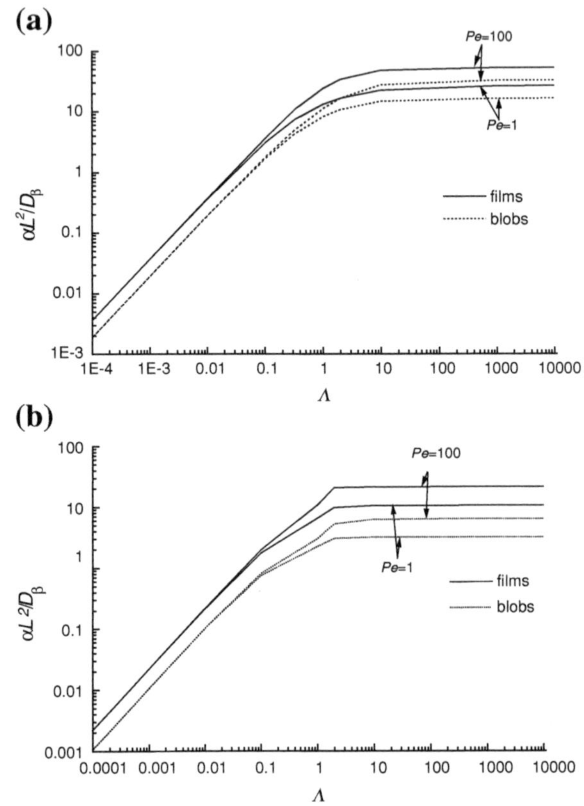

Fig. 7.37 The effect of Peclet number on the curve of the mass transfer coefficient as a function of Λ. Two values are shown, one low (Pe = 1) and another high (Pe = 100). Three dimensional unit cells with volume fractions of (**a**) $\varepsilon_\beta = 0.31$ and $\varepsilon_\gamma = 0.12$, and (**b**) $\varepsilon_\beta = 0.39$ and $\varepsilon_\gamma = 0.04$

In both cases the "aqueous" phase (β-phase) flows under Stokes flow conditions. Initially, the case of infinitely fast diffusion in the γ-phase $\left(\delta = \dfrac{D_\gamma}{D_\beta} \rightarrow \infty\right)$ is considered. The effect of the pore-scale Peclet number on the dimensionless mass transfer coefficient is shown in Fig. 7.36 for $\delta \rightarrow \infty$.

The mass transfer coefficient is nearly constant at low Peclet numbers (of the order of less than unity), and starts increasing when the Peclet number increases. The wetting condition affects the value of the mass transfer coefficient: the film (wetting) configuration of the immobile phase yields mass transfer coefficients always larger than those corresponding to blobs (non-wetting). This is a result of the higher surface area per unit volume available in films compared to those of blobs for the same volumetric fractions. In addition, it should be noted that these results are in excellent agreement with the theoretical investigation of Ahmadi et al. [6].

In Fig. 7.37, the effect of the dimensionless parameter Λ on the mass transfer coefficient is shown for two different values of Peclet number. Again, the mass transport coefficient increases linearly with Λ at low Λ values, and reaches a constant value at high Λ. As before, the mass transfer coefficient is higher when the immobile phase is wetting (films) rather than non-wetting (blobs). The coefficient is independent of the Peclet number for low Λ values. When ε_β and ε_γ vary, the resulting trends are qualitatively similar, as observed by comparing Fig. 7.37a, b. It is important to note that in these simulations, the dry porosity remains constant and equal to 0.43, ε_γ varied from 0.12 to 0.4 and, consequently, ε_β varied from 0.31 to 0.39. It is evident that as ε_γ decreases so does the mass transfer coefficient.

References

1. Achenbach, E., Reinsche, E.: Methane/steam reforming kinetics for solid oxide fuel cells. J. Power Sources **52**, 238–299 (1994)
2. Adler, P.M., Jacquin, C.J., Quiblier, J.A.: Flow in simulated porous media. Int. J. Multiphase Flow **16**, 691–712 (1990)
3. Agterof, W.G.M., Vaessen, G.E.J., Haagh, G.A.A.V., Klahn, J.K., Janssen, J.J.M.: Prediction of emulsion particle sizes using a computational fluid dynamics approach. Colloid Surf. B **31**, 141–148 (2003)
4. Aguiar, P., Adjimana, C.S., Brandona, N.P.: Anode-supported intermediate temperature direct internal reforming solid oxide fuel cell. I: model-based steady-state performance. J. Power Sources **138**, 120–136 (2004)
5. Ahmadi, A., Quintard, M., Whitaker, S.: Transport in chemically and mechanically heterogeneous porous media V: two-equation model for solute transport with adsorption. Adv. Water Resour. **22**, 59–86 (1998)
6. Ahmadi, A., Aigueperse, A., Quintard, M.: Calculation of the effective properties describing active dispersion in porous media: from simple to complex porous media. Adv. Water Resour. **24**, 423–438 (2001)
7. Ahmed, K., Foger, K.: Kinetics of internal steam reforming of ethane on Ni/YSZ based anodes for solid oxide fuel cells. Catal. Today **63**, 479–487 (2000)
8. Alves, M.A., Delgado, J.M.P.Q., Guedes de Carvalho, J.R.F.: Mass transfer from cylinders and plane surfaces buried in packed beds in alignment with the flow direction. Chem. Eng. Sci. **61**, 1174–1183 (2006)
9. Bekri, S., Thovert, J.F., Adler, P.M.: Dissolution and deposition in fractures. Eng. Geol. **48**, 283–308 (1997)
10. Bekri, S., Adler, P.M.: Dispersion in multiphase flow through porous media. Int. J. Multiphase Flow **28**, 665–697 (2002)

11. Bird, R.B., Stewart, W., Lightfoot, E.N.: Transport Phenomena. Wiley, New York (1960)
12. Bomberg, M.: Moisture flow through porous building materials. In: Report no. 52, Division of Building Technology, Lund Institute of Technology, Lund, Sweden (1974)
13. Bravo, M.C., Araujoc, M., Lago, M.: Pore network modeling of two-phase flow in a liquid-(disconnected) gas system. Physica A **375**, 1–17 (2007)
14. Carbonell, R.G., Whitaker, S.: Heat and mass transfer in porous media. In: Bear, J., Carpacioglu, M.Y. (eds.) Fundamentals of Transport Phenomena in Porous Media. Martinus Nijhoff Publ, Dordrecht (1984)
15. Chesters, A.K.: The modeling of coalescence processes in fluid-liquid dispersions. Chem. Eng. Res. Des. **69**, 259–270 (1991)
16. Coelho, M.A.N., Guedes de Carvalho, J.R.F.: Transverse dispersion in granular beds: Part II–mass transfer from large spheres immersed in fixed or fluidised beds of small inert particles. Chem. Eng. Res. Des. **66**, 178–189 (1988)
17. Colombert, R.: L'Humidité des bâtiments anciens; causes et effets; Diagnostic et remèdes, Editions du Moniteur, Paris (1975)
18. Costa-Nunes, O., Gorte, R.J., Vohs, J.M.: Comparison of the performance of Cu-CeO_2-YSZ and Ni-YSZ composite SOFC anodes with H_2, CO and syngas. J. Power Sources **141**, 241–249 (2005)
19. Costamagna, P., Magistri, L., Massardo, A.F.: Design and part-load performance of a hybrid system based on a solid oxide fuel cell reactor and a micro gas turbine. J. Power Sources **96**, 352–368 (2001)
20. Coutelieris, F.A., Kainourgiakis, M.E., Stubos, A.K.: Low Peclet mass transport in assemblages of spherical particles for two different adsorption mechanisms. J. Colloid Interface Sci. **264**, 20–29 (2003)
21. Crank, J.: The mathematics of diffusion, 2nd edn. Oxford University Press, UK (1975)
22. Dagan, G., Lessoff, S.: Solute transport in heterogeneous formations of bimodal conductivity distribution: 1. Theory. Water Resour. Res. **37**, 465–472 (2001)
23. Dullien, F.A.: Porous Media: Fluid Transport and Pore Structure. Academic Press, California (1979)
24. Fenwick, D.H., Blunt, M.: Three-dimensional modeling of three phase imbibition and drainage. Adv. Water Resour. **21**, 121–143 (1998)
25. Ferguson, J.R., Fiard, J.M., Herbin, R.: Three-dimensional numerical simulation for various geometries of solid oxide fuel cells. J. Power Sources **58**, 109–122 (1996)
26. Freitas, V.P., Guimarães, A.S.: Characterization of a hygro-regulated wall base ventilation system for treatment of rising damp in historical buildings. In: Proceedings of the 2nd Nordic Symposium on Building Physics, Copenhagen, Denmark, pp. 911–919 (2008)
27. Fried, J.J., Muntzer, P., Zilliox, L.: Groundwater pollution by transfer of oil-hydrocarbons. Ground Water **17**, 586–594 (1979)
28. Geller, J.T., Hunt, J.R.: Mass transfer from non-aqueous phase organic liquids in water-saturated porous media. Water Resour. Res. **29**, 833–845 (1993)
29. Gerami, S., Pooladi-Darvish, M.: Predicting gas generation by depressurization of gas hydrates where the sharp-interface assumption is not valid. J. Pet. Sci. Eng. **56**, 146–164 (2007)
30. Goldschmidt, M.J.V., Weijers, G.G.C., Boerefijn, R., Kuipers, J.A.M.: Discrete element modelling of fluidised bed spray granulation. Powder Technol. **138**, 39–45 (2003)
31. Guedes de Carvalho, J.R.F., Delgado, J.M.P.Q., Alves, M.A.: Mass transfer between flowing fluid and sphere buried in packed bed of inerts. AIChE J. **50**, 65–74 (2004)
32. Grace, H.P.: Dispersion phenomena in high-viscosity immiscible fluid systems and application of static mixers as dispersion devices in such systems. Chem. Eng. Commun. **14**, 225–277 (1982)
33. Grgicak, C.M., Green, R.G., Giorgi, J.B.: SOFC anodes for direct oxidation of hydrogen and methane fuels containing H_2S. J. Power Sources **179**, 317–328 (2008)

34. Guimarães, A.S., Delgado, J.M.P.Q., de Freitas, V.P.: Mathematical analysis of the evaporative process of a new technological treatment of rising damp in historic buildings. Build. Environ. **45**, 2414–2420 (2010)
35. Gurau, V., Liu, H., Kakac, S.: Two-dimensional model for proton exchange membrane fuel cells. AIChE J. **44**, 2410–2422 (1998)
36. Gummerson, R.J., Hall, C., Hoff, W.D.: Water movement in porous building materials-III. A sorptivity test procedure for chemical injection damp proofing. Build. Environ. **16**, 193–199 (1981)
37. Gvirtzam, H., Paldor, N., Magaritz, M., Bachmat, Y.: Mass exchange between mobile freshwater and immobile saline water in the unsaturated zone. Water Resour. Res. **24**, 1638–1644 (1988)
38. Gwo, J.P., O'Brien, R., Jardine, P.M.: Mass transfer in structured porous media: embedding mesoscale structure and microscale hydrodynamics in a two-region model. J. Hydrol. **208**, 204–222 (1998)
39. Hall, C., Hoff, W.D.: Water transport in brick, stone and concrete. Taylor and Francis, New York (2002)
40. Hall, C., Hoff, W.D.: Rising damp: capillary rise dynamics in walls. Proc. R. Soc. A Math. Phy. **463**, 1871–1884 (2007)
41. He, H., Hill, J.M.: Carbon deposition on Ni/YSZ composites exposed to humidified methane. Appl. Catal. A Gen **317**, 284–292 (2007)
42. Heibel, A.K., Scheenen, T.W.J., Heiszwolf, J.J., van As, H., Kapteijn, F., Moulijn, J.A.: Gas and liquid phase distribution and their effect on reactor performance in the monolith film flow reactor. Chem. Eng. Sci. **56**, 5935–5944 (2001)
43. Hirschenhofer, J.H., Stauffer, D.B., Engleman, R.R., Klett, M.G.: Fuel Cell Handbook, 4th edn. Business/Technology Books, Orinda (1997)
44. Holm, A., Kunzel, H.M.: Two-dimensional transient heat and moisture simulations of rising damp with WUFI-2D. In: Proceedings of the 2nd International Conference on Building Physics, Leuven, Belgium, pp. 363–3677 (2003)
45. Hunt, J.R., Sitar, N., Udell, K.S.: Non-aqueous phase liquid transport and cleanup I: analysis of mechanisms. Water Resour. Res. **24**, 1247–1258 (1988)
46. Inamuro, T.: Lattice Boltzman methods for viscous fluid flows and for two-phase fluid flows. Fluid Dyn. Res. **38**, 641–659 (2006)
47. Irvine, J.T.S., Sauvet, A.: Improved oxidation of hydrocarbons with new electrodes in high temperature fuel cells. Fuel Cells **1**, 205–210 (2001)
48. Jamsak, W., Assabumrungrat, S., Douglas, P.L., Laosiripojana, N., Suwanwarangkul, R., Charojrochkul, S., Croiset, E.: Performance of ethanol-fuelled solid oxide fuel cells: proton and oxygen ion conductors. Chem. Eng. J. **133**, 187–194 (2007)
49. Janssen, J.J.M., Boon, A., Agterof, W.G.M.: Influence of dynamic interfacial properties on droplet break-up in simple shear flow. AIChE J. **40**, 1929–1939 (1994)
50. Jaouen, F., Lindbergh, G., Sundholm, G.: Investigation of mass-transport limitations in the solid polymer fuel cell cathode. J. Electrochem. Soc. **149**, A437–A447 (2002)
51. Kainourgiakis, M.E., Kikkinides, E.S., Stubos, A.K.: Diffusion and flow in porous domains constructed using process-based and stochastic techniques. J. Porous Mat. **9**, 141–154 (2002)
52. Kamp, A.M., Chesters, A.K., Colin, C., Fabre, J.: Bubble coalescence in turbulent flows: a mechanistic model for turbulence-induced coalescence applied to microgravity bubbly pipe flow. Int. J. Multiphase Flow **27**, 1363–1396 (2001)
53. Kay, A.L., Davies, H.N.: Calculating potential evaporation from climate model data: a source of uncertainty for hydrological climate change impacts. J. Hydrol. **358**, 221–239 (2008)
54. Kikkinides, E.S., Burganos, V.N.: Permeation properties of three-dimensional self-affine reconstructions of porous materials. Phys. Rev. E **62**, 6906–6915 (2000)
55. Kim, T., Ahn, K., Vohs, J.M., Gorte, R.J.: Deactivation of ceria-based SOFC anodes in methanol. J. Power Sources **164**, 42–48 (2007)

56. Kimball, B.A., Jackson, R.D., Reginato, R.J., Nakayama, F.S., Idso, S.B.: Comparison of field-measures and calculated soil-heat fluxes. Soil Sci. Soc. Am. Proc. **40**, 18–25 (1976)
57. Klahn, J.K., Janssen, J.J.M., Vaessen, G.E.J., de Swart, R., Agterof, W.G.M.: On the escape process during phase inversion of an emulsion. Colloid Surf. A **210**, 167–181 (2002)
58. Klein, J.M., Bultel, Y., Pons, M., Ozil, P.: Modeling of a solid oxide fuel cell by methane: analysis of carbon deposition. J. Fuel Cell Technol. **7**, 425–434 (2007)
59. Kostoglou, M., Karabelas, A.J.: Evaluation of zero order methods for simulating particle coagulation. J. Colloid Interface Sci. **163**, 420–431 (1994)
60. Kostoglou, M., Karabelas, A.J.: Induced pulsing in trickle beds-characteristics and attenuation of pulses. Chem. Eng. Sci. **60**, 6584–6595 (2005)
61. Kunzel, H.M.: Simultaneous heat and moisture transport in building components; one and two dimensional calculation using simple parameters. Dissertaion, University of Stuttgart (1995)
62. Lam, A.C., Schechter, R.S., Wade, W.H.: Mobilization of residual oil under equilibrium and non-equilibrium conditions. Soc. Petrol. Eng. J. **23**, 781–790 (1983)
63. La O', G.J., In, H.J., Crumlin, E., Barbastathis, G., Shao-Horn, Y.: Recent advances in micro devices for electrochemical energy conversion and storage. Int. J. Energy Res. **31**, 548–575 (2007)
64. Lehr, F., Millies, M., Mewes, D.: Bubble-size distributions and flow fields in bubble columns. AIChE J. **48**, 2426–2443 (2002)
65. Lessoff, S., Dagan, G.: Solute transport in heterogeneous formations of bimodal conductivity distribution: 2. Applications. Water Resour. Res. **37**, 473–480 (2001)
66. Lin, Y.B., Zhan, Z.L., Liu, J., Barnett, S.A.: Direct operation of solid oxide fuel cells with methane fuel. Solid State Ionics **176**, 1827–1835 (2005)
67. Lu, N., Li, Q., Sun, X., Khaleel, M.A.: The modeling of a standalone solid-oxide fuel cell auxiliary power unit. J. Power Sources **1661**, 938–948 (2006)
68. Luo, L.S., Girimaji, S.S.: Lattice Boltzmann model for binary mixtures. Phys. Rev. E **66**, 035301/1–035301/4 (2002)
69. Luo, L.S., Girimaji, S.S.: Theory of the lattice Boltzmann method: two-fluid model for binary mixtures. Phys. Rev. E **67**, 036302/1–036302/11 (2003)
70. Luo, L.S., Svensen, H.F.: Theoretical model for drop and bubble breakup in turbulent dispersions. AIChE **42**, 1225–1233 (1996)
71. Matsuzaki, Y., Baba, Y., Sakurai, T.: High electric conversion efficiency and electrochemical properties of anode-supported SOFCs. Solid State Ionics **174**, 81–86 (2004)
72. Mingzhe, D., Dullien, F., Jun, Z.: Characterization of water-flood saturation profile histories by the 'complete' capillary number. Transp. Porous Media **31**, 213–237 (1998)
73. Mousavi, S.M., Jafari, A., Yaghmaei, S., Vossoughi, M., Sarkomaa, P.: Computer simulation of fluid motion in a porous bed using a volume of fluid method: application in heap leaching. Miner. Eng. **19**, 1077–1083 (2006)
74. Nagel, F.P., Schildhauer, T., Biollaz, S.M.A., Wokaun, A.: Performance comparison of planar, tubular and Delta8 solid oxide fuel cells using a generalized finite volume model. J. Power Sources **184**, 143–164 (2008)
75. Newman, J.S., Tabias, C.W.: Theoretical analysis of current distribution in porous electrodes. J. Electrochem. Soc. **109**, 1183–1191 (1962)
76. Ni, M., Leung, D.Y.C., Leung, M.K.H.: Mathematical modeling of ammonia-fed solid oxide fuel cells with different electrolytes. Int. J. Hydrogen Energy **33**, 5765–5772 (2008)
77. Park, S., Cracium, R., Vohs, J.M., Gorte, R.J.: Direct oxidation of hydrocarbons in a solid oxide fuel cell I. Methane oxidation. J. Electrochem. Soc. **146**, 3603–3605 (1999)
78. Plumb, O.A., Whitaker, S.: Diffusion, adsorption and dispersion in heterogeneous porous media: the method of large scale averaging. In: Cushman, J.H. (ed.) Dynamics of Fluids in Hierarchical Porous Media. Academic Press, New York (1990)
79. Pulkrabek, W.W.: Engineering Fundamentals of the Internal Combustion Engine, 2nd edn. Pearson Prentice Hall, New Jeresey (2003)

80. Quintard, M., Whitaker, S.: One and two equation models for transient diffusion processes in two-phases systems. Adv Heat Transfer **23**, 369–464 (1993)
81. Quintard, M., Whitaker, S.: Transport in ordered and disordered porous media: volume averaged equations, closure problems and comparison with experiments. Chem. Eng. Sci. **48**, 2537–2564 (1993)
82. Quintard, M., Whitaker, S.: Convection, dispersion and interfacial transport of contaminants: homogeneous porous media. Adv. Water Res. **17**, 221–239 (1994)
83. Ramakrishna, P.A., Yang, S., Sohn, C.H.: Innovative design to improve the power density of a solid oxide fuel cell. Power Sources **158**, 378–384 (2006)
84. Riensche, E., Achenbach, E., Froning, D., Haines, M.R., Heidug, W.K., Lokurlu, A., von Andrian, S.: Clean combined-cycle SOFC power plant–cell modelling and process analysis. J. Power Sources **86**, 404–410 (2000)
85. Sheng, J.J., Hayes, R.E., Maini, B.B., Torkite, W.S.: Modelling foamy oil flow in porous media. Transp. Porous Media **35**, 227–258 (1999)
86. Singhal, S.C.: Advances in solid oxide fuel cell technology. Solid State Ionics **135**, 305–313 (2000)
87. Skelland, A.H.P.: Diffusional Mass Transfer. Wiley, New York (1974)
88. de Smedt, F., Wierenga, P.J.: A generalized solution for solute flow in soils with mobile and immobile water. Water Resour. Res. **15**, 1137–1141 (1979)
89. Štěpánek, F., Ansari, M.A.: Computer simulation of granule microstructure formation. Chem. Eng. Sci. **60**, 4019–4029 (2005)
90. Stone, H.A.: Dynamics of drop deformation and break up in viscous fluids. Annu. Rev. Fluid Mech. **26**, 65–102 (1994)
91. Sukeshini, A.M., Habibzadeh, B., Becker, B.P., Stoltz, C.A., Eichhorn, B.W., Jackson, G.S.: Electrochemical oxidation of H_2, CO and CO/H_2 mixtures on patterned Ni anodes on YSZ electrolytes. J. Electrochem. Soc. **153**, A705–A715 (2006)
92. Suwanwarangkul, R., Croiset, E., Pritzker, M.D., Fowler, M.W., Douglas, P.L., Entchev, E.: Modelling of a cathode-supported tubular solid oxide fuel cell operating with biomass-derived synthesis gas. J. Power Sources **166**, 386–399 (2007)
93. Torres, M.I.M., Freitas, V.P.: Treatment of rising damp in historical buildings: wall base ventilation. Build. Environ. **42**, 424–435 (2007)
94. Tseronis, K., Kookos, I.K., Theodoropoulos, C.: Modelling mass transport in solid oxide fuel cells anodes: a case for multicomponent dusty gas model. Chem. Eng. Sci. **63**, 5626–5638 (2008)
95. Tsipis, E.V., Kharton, V.V.: Electrode materials and reaction mechanisms in solid oxide fuel cells: a brief review II. Electrochemical behavior vs. materials science aspects. J. Solid State Electrochem. **12**, 1367–1391 (2008)
96. Vogel, T., Gerke, H., Zhang, R., Ge-nuchten, M.V.: Modeling flow and transport in a two-dimensional dual-permeability system with spatially variable hydraulic properties. J. Hydrol. **238**, 78–89 (2000)
97. Walters, K.M., Dean, A.M., Zhu, H.Y., Kee, R.J.: Homogeneous kinetics and equilibrium predictions of coking propensity in the anode channels of direct oxidation solid-oxide fuel cells using dry natural gas. J. Power Sources **123**, 182–189 (2003)
98. Wang, T., Wang, J., Jin, J.: A novel theoretical breakup kernel function of bubble/droplet in a turbulent flow. Chem. Eng. Sci. **59**, 2593–2595 (2003)
99. Watt, D., Colston, B.: Investigating the effects of humidity and salt crystallization on medieval masonry. Build. Environ. **35**, 737–749 (2000)
100. Wexler, E.J.: Analytical solutions for one-, two-, and three-dimensional solute transport in ground-water systems with uniform flow. In: U.S. Geological Survey Techniques of Water-Resources Investigations, Book 3, Chap. B7, p. 190 (1992)
101. Wieringa, J.A., van Dieren, F., Janssen, J.J.M., Agterof, W.G.M.: Droplet break-up mechanism during emulsification in colloid mills at high dispersed fraction. Inst. Chem. Eng. **74**, 554–562 (1996)
102. Whitaker, S.: Diffusion and dispersion in porous media. AIChE J. **13**, 420–427 (1967)

103. Whitaker, S.: Simultaneous heat, mass and momentum transfer in porous media: A theory of drying. Adv. Heat Transf. **13**, 119–203 (1977)
104. Xu, J., Froment, G.F.: Methane steam reforming and water-gas shift: I intrinsic kinetics. AIChE J. **35**, 88–96 (1989)
105. Zanotti, F., Carbonell, R.G.: Development of transport equations for multiphase systems I: general development for two-phase systems. Chem. Eng. Sci. **39**, 263–278 (1984)
106. Zanotti, F., Carbonell, R.G.: Development of transport equations for multiphase systems II: application to one-dimensional axi-symmetric flows of two-phases. Chem. Eng. Sci. **39**, 279–297 (1984)
107. Zhu, H., Kee, R.J., Pillai, M.R., Barnett, S.A.: Modeling electrochemical partial oxidation of methane for cogeneration of electricity and syngas in solid–oxide fuel cells. J. Power Sources **183**, 143–150 (2008)

Index

F. A. Coutelieris and J. M. P. Q. Delgado, *Transport Processes in Porous Media*,
Advanced Structured Materials 20, DOI: 10.1007/978-3-642-27910-2,

MIX
Papier aus verantwortungsvollen Quellen
Paper from responsible sources
FSC® C105338

If you have any concerns about our products,
you can contact us on
ProductSafety@springernature.com

In case Publisher is established outside the EU,
the EU authorized representative is:
Springer Nature Customer Service Center GmbH
Europaplatz 3, 69115 Heidelberg, Germany

Printed by Libri Plureos GmbH
in Hamburg, Germany